Out of the Classroom

Observations and Investigations in Astronomy

Dennis W. Dawson
Western Connecticut State University

BROOKS/COLE

™
THOMSON LEARNING

Australia • Canada • Mexico • Singapore • Spain • United Kingdom • United States

BROOKS/COLE

THOMSON LEARNING

Sponsoring Editor: Keith Dodson
Assistant Editor: Sam Subity
Production Coordinator: Stephanie Andersen
Permissions Editor: Sue Ewing
Interior Design: Bridget Smith
Technical Illustrator: Bridget Smith

Cover Design: Bridget Smith
Cover Photo: PhotoDisc
Print Buyer: Kristine Waller
Typesetting: Bridget Smith
Printing and Binding: Webcom Limited

For more information about this or any other Brooks/Cole products, contact:
BROOKS/COLE
511 Forest Lodge Road
Pacific Grove, CA 93950 USA
www.brookscole.com
1-800-423-0563 (Thomson Learning Academic Resource Center)

Printed in Canada

10   9   8   7   6   5   4   3   2   1

ISBN: 0-534-38015-8

Table of Contents

Starred (*) experiments require the use of binoculars or a small telescope.

Foreword

This laboratory manual grew out of my perception, through years of teaching introductory astronomy classes, that students were dissatisfied with the traditional approach to lab exercises, namely analyzing data or photographs that professional astronomers had obtained at large, remote observatories. The sense of detachment from the science was frustrating to many, as evidenced by their comments on course evaluations in years when I administered traditional lab exercises. If someone else has done the observations and basic analysis, students essentially are re-hashing the work to learn techniques. While this might be fine for students training to become professional astronomers, the majority of students in typical astronomy classes are not headed in that direction.

However, students taking a science class should understand the thrill of discovery, the (necessary) commitment to understanding basic concepts and basic experimental procedures, the concept of experimental uncertainty, and the discipline of drawing conclusions from the information at hand rather than from preconceptions. How does one present this rigor in a framework which cannot make available to students the large telescopes and precise equipment usually used for gathering and analyzing astronomical information?

My solution to the problem was to develop exercises around the use of equipment that students would be likely to have on hand already (such as their eyes (!), cameras, binoculars, and occasionally a small telescope), as well as simple equipment that they could construct out of common, inexpensive materials. The goal was to present them with observational experiments that could be accomplished over one day or night, or with brief observations every few nights over a longer period, yet would yield interesting results and give the students practice in experimental procedure. Equally important was the goal of experiential learning: getting the students out of the classroom and looking at the heavens.

A preliminary version of this manual, containing thirteen exercises, was developed during the summer and early fall of 1996. The current manual contains 28 experiments, including six indoor exercises for those occasional cloudy nights! I hope that this revised and expanded manual even better realizes its objectives. Have fun with it!

Dennis W. Dawson
Western Connecticut State University
Danbury, Connecticut
Autumn, 2000

Sources

Information for this manual was drawn from a wide variety of literature and Web sources, including the following:

The AAVSO Variable Star Atlas (2nd Edition); 1990, Charles E. Scovil; Sky Publishing Corp., Cambridge, Massachusetts. [www.aavso.org]

Astronomical Tables of the Sun, Moon and Planets (2nd Edition); 1995, Jean Meeus; Willmann-Bell, Inc., Richmond, Virginia. [www.willbell.com]

Astronomy (periodical); Kalmbach Publishing, Waukesha, Wisconsin.
[www.astronomy.com]

Atlas of the Moon; 1996, Antonin Rukl; Kalmbach Publishing, Waukesha,Wisconsin.

Budapest, Mitteilungen der Sternwarte, No.76, 1980; L. Szabados (data for VV Cas)

Burnham's Celestial Handbook, Volumes One through Three; 1978, Robert Burnham, Jr.; Dover Publications, New York.

http://nssdc.gsfc.nasa.gov/photo_gallery/photogallery-jupiter.html (photo of Jupiter).

Monthly Notices of the Royal Astronomical Society, 52, 1971; S.B. Parsons and G.D. Bouw, p.138 et. seq. (data for Properties of Typical Cepheids).

Netscape Virtuoso (2nd Edition); 1996, Elissa Keeler and Robert Miller; MIS Press, New York. [www.mispress.com]

Norton's 2000.0 Star Atlas and Reference Handbook; 1991, Arthur P. Norton (18th Edition, ed. Ian Ridpath); Longman Scientific and Technical, New York.

The Observer's Handbook (Royal Astronomical Society of Canada; periodical; ed. Roy L. Bishop); University of Toronto Press, Toronto. [www.rasc.ca]

The Old Farmer's Almanac (periodical); Robert B. Thomas; The Old Farmer's Almanac, Dublin, New Hampshire. [www.almanac.com]

Science: Its History and Development among the World's Cultures; 1982, Colin A. Ronan; Facts on File Publications, New York.

Sky & Telescope (periodical); Sky Publishing, Cambridge, Massachusetts.
[www.skypub.com]

<u>Sky Catalogue 2000.0</u>; 1985, eds. Alan Hirshfeld and Roger W. Sinnott; Sky Publishing.

STAR AND PLANET LOCATOR (planisphere); Edmund Scientific Company; Barrington, New Jersey.   [www.edsci.com]

Based upon <u>Sterne and Stern Systeme</u>, (2nd Edition); 1950, W. Becker, Theodor Steinkopff, Dresden (Figure 2 on p.157).

VOYAGER II and VOYAGER III; Carina Software, San Ramon, California.
                                                                    [www.carinasoft.com]

Acknowledgments

This manual would not exist without the help of many people, not least of whom are the many students who tested the early versions of the lab experiments. But I would also like to thank the following individuals who contributed materially to its success: Bridget Smith, for her wonderful graphics skills (and patience with the author); Robert Gendler, for kind permission to use his CCD image of the Full Moon; Gary Carlson, who first saw the manual's potential; Samuel Subity, who has been a discerning editor and a good friend; Ronald Angione of San Diego State University, where the digital stellar spectra were obtained by the author, who generously provided observing time; Michael Seeds and Phillip Lu, from whose ideas the current versions of the Spiral Structure and H-R Diagram exercises were developed; and, most especially, my wonderful mother Ruth and my beautiful wife Noreen, who gave me understanding, love, inspiration, and motivation as this project came to fruition. I could not have done it without them. Many thanks to you all!

Reference Guide to Experimental Methods

Introduction to Scientific Method

Scientific method is a sort of "hyper-organized common sense" which is used for the gathering and interpretation of information. Use of the method dates back to the time of Leonardo Da Vinci, and Galileo was one of its greatest supporters. While you do not HAVE to use this method in doing scientific research (guesswork, insight and yes, even personality come into play!), you will find it a very efficient and reliable approach with the additional advantage that your results can be sometimes corroborated by others if they also follow the method.

Steps in Scientific Method:

(1) *observe* some event or condition;

(2) *explain* what you have observed (develop an hypothesis);

(3) use your explanation to *predict* some future event or condition;

(4) *observe the predicted event or condition*, comparing when /where/ how it occurs with the circumstances you predicted;

(5) if necessary (if your prediction does not exactly match what you actually see), *revise* your explanation; and

(6) with the knowledge thus gained, *go back to step (1)* and continue the process, *generalizing* the validity of your explanation for more situations.

The most general and wide-ranging explanations of physical phenomena are called **theories**; for example, the theory of gravitation and the theory of relativity.

Occam's Razor

Also implicit in the use of scientific method is the idea called *Occam's Razor,* first proposed by William of Occam (or Ockham) in the 1300s; namely, that if you have two or more competing explanations of a situation, the one which makes the fewest assumptions is the one to prefer: "simpler explanations are better ones."

Doing Scientific Experiments

One of the important aspects about doing scientific experiments is that the conditions under which they are done should be REPRODUCIBLE as much as possible. This doesn't mean that someone else can go out and observe exactly the same objects you do (after all, the heavens are constantly changing); instead, what it means is that you must be as specific as possible about HOW you performed the experiment and what equipment was used, so that someone *doing a similar experiment with similar equipment could achieve similar ACCURACY* to what you accomplished.

Steps in Making Your Experiment Reproducible:

* Provide sufficient information in the lab report about your data-gathering techniques and observing equipment, so that another person could reproduce your experiment.

* Specify the experimental VARIABLES and CONTROLS.

* Determine the level of uncertainty in your measurements (*see below*).

Experimental Variables and Controls

Variables

Variables are quantities that CHANGE during an experiment. They come in two types: independent variables and dependent variables.

The *independent variables* in an experiment are conditions **that the experimenter can deliberately alter or choose to examine when they have different values**, so as to measure the effects of the changes on other quantities. Any other quantities which change as a <u>result</u> of this alteration or choosing are called *dependent variables*.

For example, a chemist might adjust the temperature of a particular solution to investigate how the reaction rate depends upon temperature; the reaction rate therefore has become a dependent variable (its value depends upon the temperature), while temperature (the value of which was changed by the experimenter) is an independent variable.

When making astronomical observations, you might choose to alter your viewing time and/or location, investigating the consequences of doing so with regard (say) to where a constellation was seen, or whether or not a planet was visible.

Controls

It is important not only to be able to ALTER certain parts of an experiment but also to maintain some <u>unchanging</u> part, often for comparison purposes. An *experimental control* is something that is intended to remain CONSTANT while the experiment is running. Of all the possible sources of uncertainty in an experiment, you then know that the factor being controlled WILL NOT change and therefore will not make the level of uncertainty worse. A control is an important part of an experiment that should be set up before proceeding.

Examples of Variables and Controls:

Example 1:

You want to estimate how faint a star can be seen with the naked eye under different conditions, such as from a location in the countryside versus one near the center of a large city.

LOCATION therefore becomes an independent variable in your experiment. The brightness of the faintest star seen in a given location is the dependent variable.

Some controls are the part of the sky viewed (the same star pattern viewed each time), the altitude of that star pattern (have it overhead, say, instead of rising sometimes or setting sometimes), and the weather conditions (choose only clear nights); since the human eye takes about ten minutes to adapt (become sensitive) to low light levels, another control might be to remain in a dark room for ten minutes each time before starting any observations.

Example 2:

You are recording the night-to-night motion of a bright planet such as Jupiter through the background patterns of stars. You are taking still photographs (time exposures) of that part of the sky using a camera mounted on a tripod.

An independent variable in your experiment is the DATE (you can choose which nights to photograph). How the planet's position changes from one night to the next is a dependent variable.

For controls: you use the same camera and lens every night so all pictures have the same scale; you expose for the same length of time each night and use the same film type so all pictures are exposed about equally well; and you take the photographs around the same TIME each night, so that the star pattern is seen at the same altitude in the sky as on previous nights.

Experimental Uncertainties

EVERY experiment you do involves a measurement or an estimate of some sort, whether it is in determining a length or an angle or judging an amount of elapsed time. Each measurement, therefore, is NOT perfectly accurate: it contains a certain amount of UNCERTAINTY. Estimating the level of uncertainty is one of the most important tasks in successfully completing an experiment; it is a way of establishing confidence in your results.

The words "uncertainty" and "error" are used interchangeably. There are different kinds of errors, of which you should be aware.

Types of Errors:

Gross errors, commonly called "mistakes," are ones which I hope you will make a personal effort to minimize! They consist of misreading where a decimal point is, confusing centimeters for millimeters, and the like (**Duh**). One way to avoid them is to be constantly on the lookout for them! Practice with experimental techniques and number handling will also help.

Accidental errors are the typical errors of measurement. They are generally assumed to be RANDOM; that is, when you make a measurement of something, it can be a little too high or a little too low but there is really no telling which it will be on any particular occasion. [You *want* your uncertainties to be of this type (reasons below) but as **small** as practically possible.]

A worse situation is when your measurements are <u>always</u> a little high or <u>always</u> a little low but you don't KNOW it, due to "bad eyes," misadjusted or misused equipment, or poor measurement technique (such as measuring the altitudes of objects while standing on the slope of a hill rather than on level ground). Such errors are called *systematic errors*, and I hope you don't encounter them! They cannot be minimized as easily as accidental errors can (*see the section on averaging*).

Significant Figures and Measurement Accuracy

Pocket calculators and personal computers treat all numbers used for calculations as being **very accurate**: usually from ten to sixteen-digit accuracy. ***This is usually NOT true for numbers obtained through measurements***, because such numbers are more limited in accuracy by your measuring technique and the equipment you use to make the measurements.

If you measure the length of a line with a metric (millimeter) ruler, the smallest marks on it are millimeter marks. Since you can easily count the number of such marks that make up the length of the line, the number of WHOLE millimeters of its length is considered CERTAIN.

However, it is not too hard to *estimate* <u>between</u> two millimeter marks, either; say, to specify 0.2 mm or 0.5 mm or 0.8 mm. Then the length of the line is MORE accurately known — say, as 25.3 mm instead of just 25 mm — but the last number is an ESTIMATE and is therefore LESS certain. It is virtually impossible for the eye to estimate lengths to 1/100 mm (this would involve being able to visualize the distance between one mm mark and the next as being divided into 100 parts!), so in this case to express the line's length as, say,

25.32 mm is **not justified** because you can't measure it that accurately with the available equipment. The particular measuring tools and the technique of using them to obtain the measurements set a LIMIT on the number of digits which have SIGNIFICANCE in what we report. In the example above, 25.3 mm is the correct way to report the result; the number has THREE significant digits, with the first two being considered CERTAIN and the last one (the decimal part) considered REASONABLY certain.

> In general, *you can do ONE DECIMAL PLACE BETTER than the SMALLEST subdivision on your measuring instrument.* If your wall thermometer is divided into 1-degree (Fahrenheit) intervals, you can certainly read the whole numbers of degrees and can estimate between them, too, if you're careful, such as estimating 71.4 degrees instead of just 71 degrees or just 72 degrees.

You MUST keep track of significant digits in your numbers when they involve measurements, because often measurements are *combined* in different ways to yield some further information. A simple example would be multiplying the two measured lengths of the sides of a rectangle to compute its area.

> The rule of thumb for using significant figures in calculations is that *the result cannot have more significant figures than the LEAST significant number entering the calculation.*

Example of the Use of Significant Figures

Let's compute the area of a triangle from measurements of its base and height. The formula for the area of any triangle is   area = (base x height) / 2   where "2" here is a pure number (2.0000000000....) but the base and height will be measurements.

Suppose base = 21.5 mm and height = 41.4 mm; then your calculator tells you that the area is (21.5 x 41.4) / 2 or 445.05 square mm, but that last number has FIVE digits whereas the numbers you put in only have THREE. *The result must be rounded to three digits because measurements of limited accuracy were involved in getting the result.* So: our ESTIMATE of the area of this triangle is 445. square mm, and not 445.05 as the calculator would have us believe.

Expressing Uncertainty

How do you express a level of uncertainty in a measurement? The usual way is to list the value of the quantity measured (such as a length) and follow it by a "plus-or-minus" number,

e.g.,   10.2 mm ± 0.2 mm

Measurements and Deviations

What does the " ± " number signify, and how do you obtain it? It is generally a good idea to take multiple measurements of any quantity, in hopes of improving the accuracy of your final results, and this happens because we are assuming that the errors of measurement are the <u>accidental</u> (random) kind. Suppose, for example, that you are using a ruler to measure the length of a line in millimeters; let us say, further, that somehow it is <u>known</u> that the TRUE length of that line is 91.5 mm. Your measured values will usually be higher or lower than that number (though occasionally a measurement might equal that value, too). Therefore, when you listed a measurement as 91.8 mm, you would KNOW that it was 0.3 mm too high. In the table below, then, the first column lists the actual measurements while the second column lists by how much higher or lower than the ACTUAL value of 91.5 mm each measurement is. The numbers in the second column are called the DEVIATIONS from the true value.

Measurement (mm)	Deviation (mm)
91.4	-0.1
91.3	-0.2
91.8	+0.3
91.5	0.0
91.6	+0.1
91.2	-0.3
91.8	+0.3

The Usefulness of Averaging

Now add the numbers in the first column together and divide the total by the number of values (7); that is, take an AVERAGE. The average, here, is (91.4 + 91.3 + 91.8 + 91.5 + 91.6 + 91.2 + 91.7) / 7 , which equals 91.51 mm, very close to the TRUE value. More important, if you average the *deviations* in the second column, **that average is nearly ZERO.**

> IF your errors of measurement are accidental errors, **averaging <u>many</u> measurements will tend to <u>cancel out</u> the errors**. The average will represent a MORE ACCURATE ESTIMATE of the true value of the quantity being measured, and that will be true *whether or not you know what its true value is* (as long as your measurement errors really are random ones).

The " ± " number represents a sort of "average uncertainty" in your measurements and therefore must, in the example above, be related somehow to the numbers in the second column of the table. However, a straight average will NOT do here, because you've just seen that the deviations <u>average</u> nearly to zero but their *individual values* are larger.

Average Deviation

If we add the numbers <u>without regard for their sign</u> and average <u>that</u> result, it is called the *average deviation*. In the case above, the average deviation would be

$$(0.1 + 0.2 + 0.3 + 0.0 + 0.1 + 0.3 + 0.3) / 7 \quad \text{or} \quad 0.18 \text{ mm}^*.$$

This number more accurately represents the <u>range</u> of deviations seen in the table. The quantity as determined from the average of the seven measurements would have its value expressed as 91.51 mm ± 0.18 mm* .

Standard Deviation (σ)

A more accurate and meaningful (but slightly harder to compute) error number can be gotten by SQUARING the deviations, adding up the squared values, dividing by the number of data points to get a sort of "average square" value, then taking the square root of that to get what is called the **standard deviation**. This number is represented by the Greek letter "sigma" (σ)**.

In the example above,

σ = square root of [($(-0.1)^2 + (-0.2)^2 + (0.3)^2 + (0.0)^2 + (0.1)^2 + (-0.3)^2 + (0.3)^2$) / 7)]
   = 0.22 mm***

and the final result is expressed as   91.51 mm ± 0.22 mm.

There doesn't seem to be a big difference between the average deviation and σ, does there? And σ is a number which is harder to compute, so why do it? The answer is that σ is a much more meaningful way of expressing the uncertainty as long as the errors of measurement are *accidental* (random) ones.

> The number σ is a very important one in accidental (random) errors because the "odds" of any of the measurements, *including any future measurements you might make*, being **within the [average value ± σ] range** is very well known: it is about 68/100 (or 68%). The odds of any measured value (including any future one) being **within the [average value ± 2 σ] range** is around 95/100 (or 95%).

---

*   Note that, in all the above examples, we have somehow squeezed out one extra significant digit for the result! In general, averaging around ten measurements of a quantity allows you to do this because of the (hopeful) cancellation of some of the + and - uncertainties in the measurements. This is another reason for making <u>many</u> measurements of what you are observing.

**   Sometimes the standard deviation is defined as the square root of [(the sum of the squares of the deviations) / (N-1)], where N is the number of data points; in the above example, this would be the square root of [($(-0.1)^2 + (-0.2)^2 + (0.3)^2 + (0.0)^2 + (0.1)^2 + (-0.3)^2 + (0.3)^2$) / 6)]. For the purposes of this set of exercises, using N-1 seems an unnecessary complication. Dividing by N is a little more comprehensible because it is easier then to see that σ is related to an average of SQUARES of numbers.

***   See the first note!

Reference Guide to Experimental Methods

Basic Angles and Trigonometry

Astronomers usually cannot measure the actual sizes or distances between the objects they observe; instead, they must deal with *apparent* or *angular sizes* and *distances*. The <u>angular diameter</u> of an object, such as the Full Moon, is the <u>difference in direction between one edge and the opposite edge</u>, usually measured in degrees. The Full Moon's angular diameter is about half a degree.

You are probably familiar enough with degree measurement in general: a right angle is 90 degrees, each vertex (point) of an equilateral triangle is 60 degrees, and so on. *Figure 1* may refresh your memory!

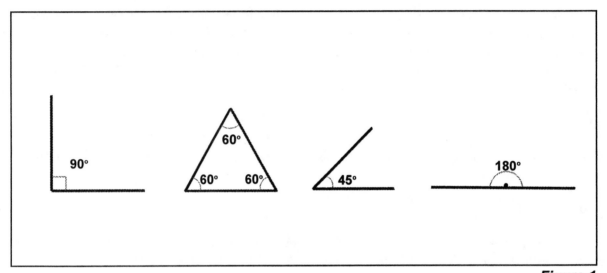

Figure 1

Astronomers also use smaller angle divisions called *minutes of arc* (') and *seconds of arc* ("). There are 60 minutes of arc (= 60') in 1 degree and 60 seconds of arc (= 60") in 1 minute of arc (or 3600 seconds of arc in 1 degree). One second of arc is a very small angle! But it is typical of the apparent diameters of stars as seen through Earth-based telescopes.*

The Small-Angle Approximation

It was Aristarchus of Samos (310-230 B.C.) who first developed a version of a fundamental formula of geometry that could be applied to astronomical observations. The relationship, called the **small-angle approximation**, is most accurate for objects with angular

---

*   <u>Note:</u> The apparent diameters of stars seen through Earth-based telescopes are NOT measurements of the stars' ACTUAL diameters. Most stars are so distant that their edges cannot be readily seen in any telescope! But Earth's atmospheric motions (winds) BLUR the tiny images of stars into much larger disks, with angular diameters around 1" under the best conditions, such as atop a high-altitude observatory. From space telescopes, stars' images are much sharper, but it is still exceptionally difficult to see disks for all but the nearer, supergiant stars.

diameters less than a few degrees, and it can be conveniently stated as:

$$\frac{\text{True diameter, D}}{\text{True distance, d}} \quad = \quad \frac{\text{Angular (apparent) diameter (”)}}{206,265 \text{ ”}}$$

The small-angle approximation says that the proportion between an object's true size and its distance from the observer is the SAME proportion as its angular diameter has to the number 206,265". Diameter and distance must both be in the same units (such as miles or kilometers). See *Figure 2*.

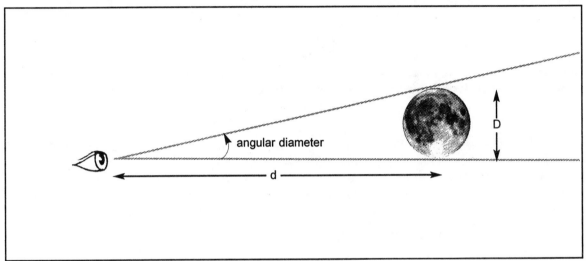

Figure 2

An example of the use of the small-angle approximation is the calculation of the actual diameter of the Sun from its apparent (angular) diameter of 32' (= 1920" or a bit more than half a degree). The Sun's average distance from Earth is 149,600,000 kilometers, so we have that

$$\frac{\text{Sun's true diameter (km)}}{149,600,000 \text{ km}} \quad = \quad \frac{1920\text{”}}{206,265\text{”}}$$

The righthand proportion evaluates as 0.009308 (expressed to four significant figures, the same as 1920" has), so our estimate of the Sun's true diameter is

$$D = 0.009308 \times 149,600,000 \text{ km} = 1,392,000 \text{ km} \ .$$

The proportion 1920" / 206,265" is roughly 1 / 107; another way to look at the example above is to say that when the Sun is viewed *from a distance equal to 107 of its own diameters*, its angular size is about half a degree.

Given any two of the three quantities in the formula, you can use the small-angle approximation to find the third. It's a very handy equation!

Trigonometry of Right Triangles

Trigonometry (literally, the measurement of triangles) is the set of rules which relate the lengths of sides of triangles to the angles between those sides. The rules are simplest when you consider right triangles: those with one internal angle of 90 degrees.

See *Figure 3a* below. The lengths of the sides which make the right angle are traditionally called "a" and "b". The remaining side (opposite the right angle), of length "c", is called the **hypotenuse**. As you probably know, the lengths of the three sides are related through the Pythagorean theorem, namely

$$a^2 + b^2 = c^2$$

The figure shows a right triangle with the lengths of the sides (a:b:c) in the proportions 3:4:5. This certainly follows the Pythagorean theorem, since $3^2 + 4^2 = 5^2$.

Figure 3b shows how the OTHER angles in a right triangle are defined by the lengths of the triangle's sides. In both triangles, the length "a" is the same, but in *Figure 3a* the angle (AC) between sides a and c is larger than it is in the righthand triangle. At the same time, the angle (BC) between sides b and c is SMALLER in the lefthand triangle than it is in the righthand one.

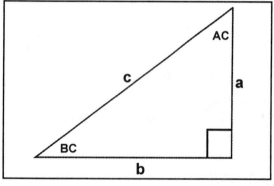
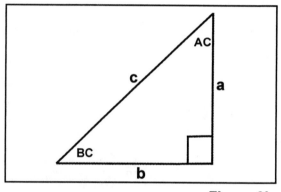

Figure 3a   **Figure 3b**

The proportions between lengths of triangle sides are related to the angles in the triangles. The values of the proportions make up what are called the **sine**, **cosine** or **tangent** values of the particular angle. Those "trig functions" are defined as:

Sine of the angle =   Length of side opposite the angle
                         Length of hypotenuse

Cosine of the angle =   Length of side adjacent to the angle
                         Length of hypotenuse

Tangent of the angle =   Length of side opposite the angle
                         Length of side adjacent to the angle

In the triangles in *Figures 3a* and *3b*,

$$\sin AC = b / c \qquad \cos AC = a / c \qquad \tan AC = b / a$$

and

$$\sin BC = a / c \qquad \cos BC = b / c \qquad \tan BC = a / b$$

To get from the sine, cosine or tangent of an angle to the angle itself, a *mathematical operation* must be performed upon that function. Doing that operation is called "taking the inverse" or "taking the arc-(name)" of the function, such as inverse tangent or arc-tangent. If all this seems like gobbledygook, don't worry! You can do it easily on any calculator with trig function keys and a "second function" or "inverse" key.

Suppose, for example, that you measured a = 7.1 mm, b = 6.5 mm and c = 9.6 mm. Then (for example) sin AC = b / c = 6.5 mm / 9.6 mm = 0.68 . On your calculator, look for "sin^{-1}" above your SIN key. If you find it, that operation is reached by pressing the "2nd function" or "2nd" key first, then the SIN key. Otherwise, look for an INV key and press it, then the SIN key. Either way, the angle you want is

$$AC = INV \ SIN \ or \ \sin^{-1} \ 0.68 = 43. \ degrees$$

Sines, cosines and tangents are useful things to know! One quick way to appreciate them is to draw right triangles of the same base length but different heights on graph paper, *measure* the internal angles with a protractor, then see how the ratios of the sides relate to those numbers.

Some useful values of sine, cosine and tangent:

$$\sin 30° = 0.50 = \cos 60° \qquad \sin 45° = \frac{\sqrt{2}}{2} = 0.707 = \cos 45°$$

$$\sin 60° = \frac{\sqrt{3}}{2} = 0.866 = \cos 30° \qquad \tan 45° = 1$$

See *Figure 4*. Suppose you knew the height H of a radio tower and wanted to know how far from it (d) you were standing. You would measure the *angular* height (A°) of the tower and use:

$$\tan A° = H / d \qquad or \qquad \mathbf{d = H / \tan A°.}$$

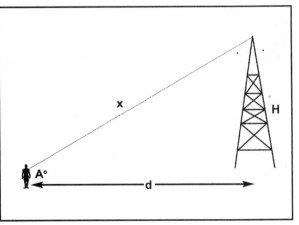

If you wanted to know the distance x to the TOP of the tower, you would use:

$$\sin A° = H / x \qquad or \qquad \mathbf{x = H / \sin A°.}$$

(Of course, if you knew BOTH H and d, then $x^2 = H^2 + d^2$ and you wouldn't need the sine or tangent!)

Figure 4

Basic Graphing

Often in scientific work, we need to compare one measured quantity against another (an "x" versus a "y"). This can be done by simply listing the values together as different columns of a table, but an even better method is to display them in a graph. The eye and brain are particularly adept at analyzing information when it is displayed in this way. A graph can quickly tell us how closely the two quantities are RELATED to each other (does the value of one <u>depend</u> upon the value of the other, and under what conditions?) and whether that relationship is a straight-line (proportional) one or is more complicated.

The guidelines below are useful for ALL graphing you may do:

(1)   Your graph should always have a TITLE (at the top of the graph itself) which briefly explains what is being plotted.

(2)   Each graph axis should be labeled with the name of the quantity being plotted on that axis, and (in parentheses) the UNITS in which it is expressed.

(3)   The graph should fill MOST of the page, with reasonable allowance for numbering and labeling along the axes and room for the title at the top. To that end, you need to look at the values of your data points and decide:

>   (a) at what values to start and end each axis, <u>and</u>

>   (b) into what values you will subdivide each axis. The subdivisions are particularly important (*see below*).

A Sample Graph

As an example, let's consider a graph of the air temperature and pressure (as measured at our local weather station) during the afternoon of July 21. The values were measured at one-hour intervals, starting at noon of that day. The data are shown in the table that follows:

Temperature (Fahrenheit)	Pressure (millibars)	Time (EDT)
74.5	1011.5	noon
76.7	1012.8	1 p.m.
78.2	1013.2	2 p.m.
81.4	1013.7	3 p.m.
82.9	1014.0	4 p.m.
81.8	1014.3	5 p.m.
81.1	1014.2	6 p.m.
80.5	1014.0	7 p.m.
79.9	1013.6	8 p.m.
77.6	1013.2	9 p.m.

The first thing to do is to establish the RANGES of numbers to be shown on the graph. The temperature (which we'll plot as "x," the independent variable) ranges from 74.5 degrees to 82.9 degrees, while the pressure (which we plot as "y," the dependent variable) ranges from 1011.5 mb to 1014.3 mb. { Please note that *neither of these ranges starts anywhere near zero, and that it would be pointless to begin the graph axes at values that low!* Graphs do NOT all have to start at (0,0)! }

The second thing to do is to decide upon how many divisions each axis will have. It is most useful (because you will sometimes have to work between the divisions when reading the graph) to have a multiple of 1, 2, 4, 5 or 10 squares as the number of divisions, and not at ALL useful to have multiples of 3, 6 or 7 squares!

For the temperature (x) axis, choose a starting value BELOW 74.5 and an ending value ABOVE 82.9 degrees. If the starting value were 74 degrees and the ending value were 83 degrees, there would be 9 whole-degree divisions, which is acceptable. For the pressure (y) axis, 1011.0 could be the starting value and 1015.0 the ending value, with 1.0 mb subdivisions. Let's look at *Figure 5*:

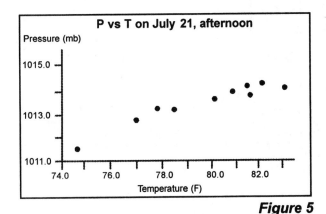

Figure 5

It contains quite a bit of information, the most obvious part of which is that the two quantities being plotted are RELATED to each other: when one is high, so is the other. A next step might be to express the relation between the points by FITTING A LINE through them; this is shown in *Figure 6*:

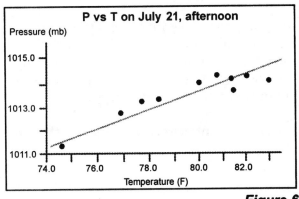

Figure 6

Reference Guide to Experimental Methods

Trend Lines

The line itself is sometimes called a <u>trend line</u> because it shows the sense (direction) of the relationship. Notice that it "threads" its way among the points so that about an equal number of points lie above the line as below it (occasionally, some points may also lie **on** the line).

Scatter

How far the plotted points SCATTER above or below the line is a measure of our confidence in that line actually expressing the relationship between the two quantities; *Figure 6* shows a good fit with small scatter (σ), which is what we want! Compare this with the next graph, *Figure 7*, which shows a (still good!) fit to points which have a LARGE scatter; the line DOES thread accurately through the points, but because there is so much scatter we don't have as much *confidence* in the *trend*.

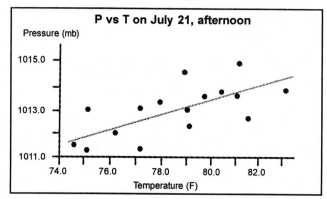

Figure 7

Rejecting Data Points

Sometimes a question arises about whether a particular data point on a graph is accurate, as the point in question seems to lie noticeably outside the TREND suggested by looking at the other points. There are rules which let you reject points from your data, but you should NOT do this very often and you should have VERY definite reasons for thinking the point is bad, other than "it doesn't fit the trend!" Sometimes, the point you rejected proves to be an important one, showing a whole NEW trend! *Reject points with caution!*

Figures 8a and *8b* illustrate the ideas involved. On *Figure 8a*, the indicated point lies well above the trend line already found but also somewhat to its right. Should it be rejected, or does it show that the relationship is CURVED toward the upper right? (The best solution, if it is still possible, is to go and get MORE data in that range of values and see whether they, too, fall in that part of the graph.)

Figure 8b is much easier to interpret, as most of the points follow a pretty obvious straight-line trend.

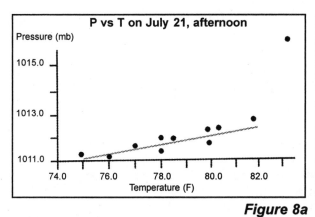

Figure 8a

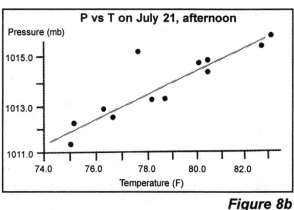

Figure 8b

If a discordant point lies more than twice the average scatter (strictly, 2σ) away from the line, there is some justification in rejecting it.

Visual Observing Techniques

There are several things to consider when setting out to make any kinds of observations with the eye, and some additional considerations when starting to do night-time observations.

For the reasons outlined earlier, you should NEVER make just ONE observation of something; make several observations and calculate their average value.

The "Near Point"

With all sighting devices that have calibration marks of some kind, those marks should be no closer than 10 inches to the eye. This distance is called the "near point" and is the closest that something can be to an average human eye and still be in focus. That is, for an average human eye, objects more than 10 inches from it (including those at "infinity," such as stars and planets) will all be in focus. If you wear corrective lenses, you will have to allow for them when making your observations (such as by removing or wearing glasses when looking at objects of different distances).

The Wonderful Human Eye!

The retina of the human eye contains two kinds of electrochemical light sensors: about 3 million <u>cones</u>, which respond to high (daytime) light levels, and about 100 million <u>rods</u>, which respond to lower (night) light levels. <u>Directly</u> at the back of the eye, in a spot called the fovea centralis, there are no rods at all; but, there are many around the <u>rest</u> of the retina, which means that at night you can get somewhat greater sensitivity by looking slightly "out of the corner" of your eye at an object; this is called *averted vision*.

The night-time (rod vision) eye has a lowered sensitivity to RED light compared to the daytime eye, which means that if you are reading a star chart, or measuring angles or lengths on a measuring device at night, you should use a RED FLASHLIGHT (not a white-light one) to help you see. (You can make a red flashlight in several ways: buy a piece of red cellophane to cover it, paint the end with red nail polish, or simply cover the end with a brown paper bag.)

Another reason to avoid looking at bright (non-red) lights at night is that such strong light will temporarily break down the chemical (*rhodopsin* or "visual purple") in the retina which allows us to have good night vision; after being "zapped" by white light (as happens, for example, when we go from a brightly-lit room into the dark), we are "night blind" for about 5 minutes, and it takes about 20 minutes for the eye to fully recover its low-light sensitivity. To get the most out of night-time observations, stay in a relatively dark location (away from direct, bright lighting) for 10-20 minutes before observing; and, if you need some light, use red light only.

Causes of a Bright Night Sky

"Light pollution" is an environmental concern caused by the wasteful use of lighting in our society. Improperly shielded lighting, as well as the over-lighting of many areas, not only wastes energy dollars but also drowns out our view of the dark sky. To see the most stars and other celestial objects, try to find a place well away from direct city lighting. The night sky can be significantly darker even a mile or so from an urban center, and much darker five to ten miles from it.

> Moonlight will also light up the night sky. The effects are worst around Full Moon.
>
> When the Moon's phase is within a few days of Last (3rd) Quarter, the sky will be dark before midnight (the Last Quarter Moon rises around midnight). The First Quarter Moon lights up the early evening skies, but not nearly as much as the Full Moon does.
>
> Consult your calendar, local paper (weather section) or almanac for the dates of the Moon's different phases. When you observe in moonlit skies, include an estimate of how close the Moon is (degrees) to the part of the sky being studied.

Other Observing Equipment

Simple Measuring Devices You Construct

For some of the exercises in this collection, you will need to construct simple devices to measure lengths and angles. These may be made of very inexpensive materials such as rulers, protractors, plastic, cardboard, styrofoam, foam board, dowel rods, and other substances available around the house or in department and homegoods stores.

The bottom line in constructing a piece of observing equipment is that it should be fairly sturdy and accurately calibrated, as you may be using it for more than one experiment. If a hand-held device wobbles, twists and slips when you use it the first time, it won't give you very accurate results.

Telescopes, Binoculars and Other Optical Devices

Some of the experiments involve the use of binoculars or small telescopes, which you may own or can borrow. Telescopes, binoculars, spotting scopes, camera lenses and other optical devices are all characterized by a number called the FOCAL LENGTH and another number called the APERTURE.

The focal length (F) of an optical device determines how LARGE objects will look when viewed through it (or photographed through it, as in the case of a camera lens).

The longer the focal length is, the more MAGNIFIED is the view. This hardly affects the appearance of the STARS (which are so distant that they still appear as points even under magnifications of hundreds of times), but it DOES enlarge the view of the Moon and other objects that have some apparent size.

The FIELD OF VIEW (how much of the sky one can observe) is also affected by focal length; longer focal length means a NARROWER field of view. So, there is always a trade-off between the magnification of some celestial object of interest and the region of sky in which it is being viewed.

The aperture (D) of an optical device (the diameter of its primary lens or mirror) is related to how BRIGHT objects will look when viewed through it.

The brightness of POINT sources of light (such as stars or distant streetlights) is in proportion to D x D. The larger the aperture of your optical device, the brighter such objects will appear.

The brightness of EXTENDED sources of light (like the Moon or a patch of night sky) is in proportion to (D x D)/(F x F). Therefore, an optical device of large aperture but short focal length will be best for making extended light sources look bright.

Binocular Particulars

Binoculars are often described by expressions like "7x50" or "9x21" (which are read as "seven exx fifty" or "nine exx twenty-one"). The "7x" or "9x" notation refers to the binocular's built-in _magnification_ (7 or 9 power: increasing the apparent size of extended objects seven or nine times), while the number that follows (such as 50) refers to the binocular's _aperture_ (diameter of each front lens), in millimeters. [One inch is 25.4 mm, so an aperture of 50 mm means a lens about 2 inches in diameter.]

Though binoculars with 16x or 20x magnifications are sold, they are usually very large, heavy, and bulky: in short, not very portable. For general sky-scanning as well as terrestrial viewing, it is better to stick with the lower-power but lighter weight binoculars, such as 7x35 or 7x50.

Reference Guide to Experimental Methods

Sometimes, the binocular manufacturer will provide the "field of view at 1000 m" for the binoculars, such as 140 meters at 1000 meters. To convert this into degrees, simply multiply the proportion (here, 140/1000) by **57.3 degrees**.

For Example:

field of view = (140 meters / 1000 meters) x 57.3 degrees **= 8.0 degrees***

Telescope Particulars

For the kinds of experiments you will be doing, the kind of telescope or mount is relatively unimportant. What is important is that you know F and D for the telescope, and also what eyepiece you are using.

An eyepiece is a small lens (or sometimes a set of small lenses) mounted in a cylindrical metal housing. This is, in turn, put into the focusing tube of the telescope and the observer looks through the eyepiece to view celestial objects.

Since the eyepiece lens is an optical device, it also has an aperture (d) and a focal length (f). The focal length in millimeters is often stamped on the metal housing of the eyepiece.

For visual observing, the magnification of a telescope with focal length F used with an eyepiece of focal length f is $M = F / f$.

For Example:

A telescope of moderate aperture which is available commercially is the Celestar 8, manufactured by Celestron International of Torrance, California. This is a telescope with an aperture of D = 8 inches or 203 millimeters and a focal length F = 80 inches or 2032 millimeters. Among the eyepieces available for this telescope are ones with focal lengths f = 40 mm, 26 mm, 12.5 mm and 7.5 mm. What is the magnification of the telescope when used with each of the listed eyepieces?

For any eyepiece of focal length f used with the Celestar 8, the magnification will be M = F / f = 2032 mm / f . Plugging in the various values of f given above, we compute magnifications of about 51x, 78x, 162x and 271x, with the lowest magnification being when the eyepiece with f = 40 mm is used.

There are many different kinds of telescopes: *refractors*, which use lenses to form images; *reflectors*, which used curved mirrors to do the same thing; and *lens-mirror hybrids*, which have both a curved mirror and a large, corrective lens that works with it. You can see all of these varieties, and several different kinds of telescope mounts, in any good science store or while browsing through *Sky & Telescope* or *Astronomy* magazine.

---

* The binocular field of view (either eye) will be a circle looking across 8 degrees of sky.

Camera Particulars

Like any other optical systems, camera lenses have an aperture D and a focal length F. The barrels of camera lenses are stamped with the focal length; a typical value for an ordinary, moderately wide field lens is F = 50 millimeters. Lenses of much shorter focal lengths (say, 28 mm) have a much wider field of view and are called wide-angle lenses or (in the extreme cases) "fish-eye" lenses. Lenses with focal lengths much longer than 50 mm (say, 200 mm) have greater image magnifications but narrower fields of view and are usually called "telephoto" lenses.

Most camera lenses contain an adjustable opening called an IRIS. When the iris is wide open, the full aperture of the lens is usually available to gather light for photographs; but, when the iris opening is smaller, it is THAT opening which acts as the aperture of the lens.

In principle, the iris of a camera lens could be adjusted to any aperture between being fully closed and open to its largest diameter. However, in practice, irises in camera lenses can be opened only to certain sizes, called STOPS. Each stop is assigned a number known as the *f-number*. You have probably seen these on cameras (usually: 2.8, 4, 5.6, 8, 11, 16, 22) and may have wondered what they were.

> The f-number of the iris stop in a camera lens is the **ratio** of the lens **focal length** to the **diameter** of the **iris** *at that stop*. A larger f-number means an iris which is closed down more in size and letting in less light.

> The amount of light let in depends upon (D xD)/(F x F) for extended sources of light, such as landscapes or views of the night sky. In a camera lens, *F is not changeable; only D can be changed, by adjusting the size of the iris.* If the f-numbers above are all SQUARED, the results will differ from each other by roughly a factor of TWO (*e.g.*, 7.8, 16, 31, 64, 121, 256, 484). For each f-number higher in the sequence that you adjust the camera lens, you let in HALF as much light; that is, a lens set to 4 lets in half the light that it would if set to 2.8 .

For most of the experiments in this laboratory manual, you may be able to use your camera with the lens wide open (at a low f-number). However, if the sky in which you photograph is very **light polluted** (illuminated strongly by artificial lighting), it may be advisable to set the camera iris to a slightly higher f-number when exposing the film; you will lose some star images but have a darker sky background for the photograph. Alternatively, use a wide-open camera lens with a SLOW speed film (like ASA 64 or 100) which responds less to the background sky brightness.

Astrophotography (photographing stars, Moon and planets) usually requires a sturdy support for the camera (a tripod) and the capability of taking time exposures (a "time" setting on the shutter speed dial and a locking cable to hold open the shutter for long exposures). Check especially whether your camera has a time-exposure capability; many modern "do everything" cameras do NOT have this feature.

Weather Conditions

Many locations have spells of cloudy weather. While some of the exercises in this manual require observations over several days or nights, others can be accomplished with only one set of observations. In this way, some compensation can hopefully be made for the meteorological version of Murphy's Law.

Clothes for Comfortable Observing

Since you will be making many of your observations at night, some information about what to wear is appropriate. Even a warm summer night can turn chilly or clammy as the day's heat escapes into the clear night sky; observing during late autumn or early spring can be downright unpleasant unless you are prepared for the weather conditions. Also, remember that you will often be standing in one location to obtain your observations; if you are not moving around much, you will lose a significant amount of heat through the soles of your shoes and from wind cooling.

Staying warm on cool or cold nights is a matter of covering those parts of the body from which heat will escape most readily: the top of the head, the ears, the neck, the wrists and hands, the ankles, and the soles and sides of the feet. For summer evenings, keep a jacket and/or sweatshirt handy and wear rubber-soled shoes. For cooler weather, add some head protection (like a hooded parka or a pullover hat) plus a scarf, gloves, a sweater, thicker socks, and warmer (more thickly lined) shoes.

Trust the word of a long-time observer: If your extremities get too cold, observing can turn into sheer misery! ***Assume it's going to be colder than you think.*** It's much better to take along extra apparel and not use it than to get caught without enough.

Keeping an Observer's Log

It is important for you as an observer to keep a careful record of what you are doing. Careful records make for better lab reports, and they also sometimes allow you to discover the cause of problems with your data (such as misidentification of a bright star or planet) should they reveal themselves later. A sample observer's log page follows, which contains the *minimum* information you should be recording. Feel free to add other information which seems important; but, once you settle on a format for your log pages, STICK with it for all experiments.

Sample Observer's Log Page

Observer(s): _____ Date: _____

General Location (*e.g.*, city name): _____

Specific Location (*e.g.*, backyard, parking lot):_____

Name of Experiment: _____

Equipment Used for Experiment (give details):_____
_____
_____

Weather and Local Lighting Conditions: _____
_____
_____

Moon's Phase and Location (altitude, direction) in sky:_____
_____

Observations were carried out between _____ a.m./p.m. and _____ a.m./p.m.

Time Zone _____ Standard Time_____ Daylight Time _____

Object(s) observed:_____

Approximate Altitude(s) of object(s):_____

Approximate Direction(s) to object(s): (specify times when necessary)
_____

Visual sketch of object	Binocular or telescope sketch of object
(specify size of field of view, in degrees)	(specify size of field of view, in degrees)

Your Laboratory Reports

The results of all your experiments should be submitted as laboratory reports. Usually, these will not be very LONG — a few pages is typical — but they DO have to contain some specific sections. Those are described below.

The title of the exercise, your *name*, and the *date* the report is submitted are three obvious but necessary things.

A description of your equipment and observations comes next. This should include a brief summary of the <u>purpose</u> of the exercise, details about the equipment and how it was constructed, when and how observations were taken (and of what objects), the <u>variables</u> and <u>controls</u>, and what kinds of calculations (when applicable) were needed to get further results. [Do NOT present specific data here; just convey a general idea if how you did the exercise.]

Data and results should follow. These can be in table and/or graphical form; for some exercises, drawings and specific information about location, weather conditions, time of observation, etc., will be needed. An important part of your results will be evaluating their level of UNCERTAINTY, and assigning some sort of number value like an average deviation or σ when appropriate. Even when a number value cannot be found, you should have a good idea <u>where</u> the greatest difficulties in obtaining accurate information were originating in the exercise (were errors of measurement mostly to blame? was the observing location too brightly lighted? was bad weather the main concern?).

Your *conclusions* should complete the report. What concepts or relationships were demonstrated to you through the exercise? In what way(s) could the exercise be made more <u>accurate</u> or of more <u>general</u> application? [To say "use better equipment" is the "easy" way out. Be more creative than that! If you HAD to use the same equipment (there being none better), how would you proceed? Consider how making different kinds of observations or having a different experimental procedure might affect the experiment.]

Sample Lab Report

Estimating the Number of Stars Visible to the Naked Eye
Celia K. Smith
September 23, 2002

Equipment and Observations: The purposes of this exercise were (1) to estimate the number of stars visible to the eye by taking samples around the sky, and (2) to investigate how the presence of the Moon in the sky would affect this number.

I looked through a paper-towel tube whose diameter and length I measured with a millimeter ruler. I observed on two different evenings from the University campus, in the student parking lot which did not have much bright lighting nearby. Observations on both nights started around 8:00 p.m. The dates were September 8, a clear night with no Moon; and September 20, also a clear night with a nearly full Moon. On each night, I chose ten different sky directions to measure: some high, some near the horizon but none right next to the Full Moon on the second night. From my numbers, I calculated an average value for each night and also a standard deviation. Then the total number of stars in a hemisphere of the sky on each night was computed from

$$N = (8L^2 / D^2) \times (\text{average number of stars})$$

and the uncertainty in that number is

$$\text{Uncertainty} = (8L^2 / D^2) \times (\text{standard deviation})$$

As controls in this experiment, I used the same paper-towel tube each night, observed on nights when the sky seemed to be clear, observed in ten different directions each time, and observed from the same location at about the same time each night. An independent variable in the experiment was the Moon's phase. A dependent variable was the number of stars seen on each night, which depended on the Moon's phase and could also depend on where I was aiming the paper towel tube.

Data and Results. I measured the dimensions of the paper-towel tube as $L = 298.1$ mm and $D = 43.9$ mm (averages of several measurements), so $8L^2 / D^2 = 369$. The table below shows the numbers of stars counted in each of the ten areas I chose, on the two different nights.

Reference Guide to Experimental Methods

(September 8)		(September 20)	
Area	# of Stars	Area	# of Stars
1	10	1	1
2	12	2	4
3	15	3	3
4	6	4	1
5	8	5	5
6	11	6	0
7	9	7	4
8	10	8	6
9	13	9	5
10	7	10	8

Average = 10.1
Std.deviation = ± 2.6

Average = 3.7
Std. deviation = ± 2.4

N = 369 x 10.1 = 3730 stars
Uncertainty = ± 959 stars

N = 369 x 3.7 = 1370 stars
Uncertainty = ± 886 stars

Conclusions.   In this experiment, I learned how a very large number of stars can be estimated by taking samples. I learned that on a dark, moonless night a person might be able to see several thousand stars at any time. The Moon, however, greatly drops the total number of visible stars (from 3730 to 1370) by lighting up the sky and making stars harder to see.

If I were to do this experiment over, I would take more samples of the sky, in various directions and at different elevations, to get a more accurate average number of stars. I would also see how the total number of stars depends on the Moon's phase at other phases besides Full. The paper towel tube had kind of a narrow view; the experiment could be repeated with a bigger tube, which would also let me count more stars.

Reference Guide to Experimental Methods

Using a Planisphere

You Will Need: • Planisphere
• Red flashlight
• One clear night for observations

Purpose: To gain familiarity with the use of a planisphere (also called a star and planet finder)

Background:

A planisphere (literally, a flattened sphere) is a convenient way of representing the constellations visible from a particular latitude at any time of night for any night of the year. The planisphere is an aid to learning the night sky.

About two-thirds of the 88 established constellations will be visible during the year to a Northern Hemisphere observer, and these are shown on the planisphere by lines connecting the brighter stars; the very brightest stars also are named (for example, *Sirius, Vega, Arcturus*).

A planisphere does have limitations. One is that it is designed for use at locations not very far north or south of a particular latitude (typically, 40 degrees North); if you want to see the sky from another latitude, you must buy a planisphere designed specifically for it. (These are available from some manufacturers.)

Another limitation is that the positions of the Moon and planets cannot be printed on the planisphere; they are too changeable. (On the other hand, if you know in what constellation a planet is located, you can use the planisphere to tell whether, and when, that constellation will be visible; if it is, you should be able to see the planet, too.)

Figure 1 shows the design of a typical planisphere. The large dark oval is the "dome" of the sky above the observer at any time; its center is the point <u>straight</u> overhead (the **zenith**). The pivot point of the planisphere wheel (usually a rivet) is the north celestial pole, an extension of Earth's rotation axis through the planet's north pole and onto the sky.

An object AT the north celestial pole will not appear to change position at all on any night, being directly "above" Earth's rotation axis; very close to this point (the pivot point) is the moderately-bright star *Polaris*, often called the North Star or the Pole Star because of its always being found at the same location while other stars are rising and setting.

Exercise 1: Using a Planisphere

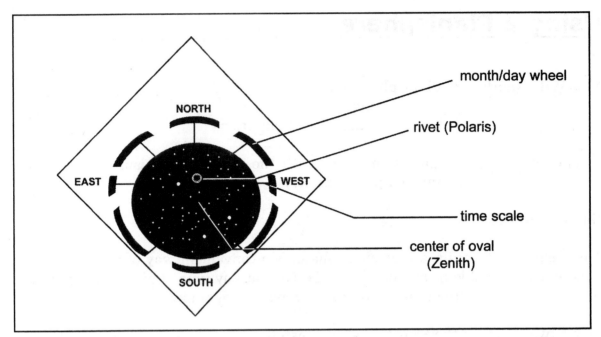

Figure 1

Also shown on planispheres are two curves, labeled EQUATOR and ECLIPTIC. The first curve is the **celestial equator**, the projection of Earth's equator onto the sky; it is the dividing curve between constellations easily visible in the Northern Hemisphere and those more easily seen from the Southern Hemisphere (some southern constellations, such as SAGITTARIUS, are visible from the Northern Hemisphere, but they are low in our skies).

The **ecliptic** is where the Sun, Moon and bright planets are found. The twelve constellations found along the ecliptic are collectively called the **zodiac**: figures of people and animals.

Procedure:

To set the planisphere for a certain date and time, choose the time of night at which you will be observing.

NOTE: All times are in Standard Time. If you are on Daylight Saving Time, *subtract 1 hour* from your chosen time to convert it to the Standard Time used by the planisphere.

Locate that time on the "time" scale which runs along the inner edge of the planisphere. Now turn the month/day wheel until the current month and date **line up** with the time mark you chose.

The planisphere now will be displaying those stars which are above the horizon on the particular month, date, and viewing time you chose. However, exactly where in

Exercise 1: Using a Planisphere

the sky these will be seen depends, of course, upon the direction toward which you are looking.

If you want to view the sky while facing **south**, HOLD THE PLANISPHERE OVER YOUR HEAD (gaze upwards at it) *with the word "South" at the bottom*. To use the planisphere while facing **east**, again hold the device over your head, but ROTATE it *until the word "West" is at the bottom*. See *Figure 2*. Remember: if you change the direction you are facing, rotate the planisphere until the new direction is at the bottom of your view.

SOUTH

WEST

Figure 2

For Example:

Set the planisphere for *10 p.m. Standard Time on July 25*. Now hold the planisphere over your head with the word "South" at the bottom (I will assume that if you were outside, you would really be facing south at this time, too!). Very near the center of the viewing oval (that is, at the zenith) is the bright star *Vega* in the constellation LYRA (the harp). Low in the southern sky is the constellation SAGITTARIUS.

Now <u>turn yourself</u> so that you are facing 90 degrees to the right (clockwise) of where you started; also, rotate the planisphere (still above your head) so that the word "West" is at the bottom. Prominent high in the western sky at this time is the bright star *Arcturus* in the constellation BOOTES (the shepherd), while much lower (near the horizon, a bit south of west) is another bright star, *Spica*, in the constellation VIRGO (the maiden).

Finally, if you turned the planisphere (and yourself) for the view facing north, you would notice the BIG DIPPER pattern in the northwestern sky, the constellation CASSIOPEIA to the northeast, and the Pole Star ("rivet") straight ahead of you and not quite halfway up to the zenith (center of the oval).

Exercise 1: Using a Planisphere

Observations:

Consult a local newspaper, an almanac like the *Old Farmer's Almanac*, a periodical like *Astronomy* or *Sky & Telescope*, a Web site or your instructor to determine in what constellations the planets Venus, Mars, Jupiter and Saturn can currently be found.

Take the planisphere outside with you on a clear night without too much moonlight (i.e., avoid nights near a Full Moon). Find a fairly dark location (away from direct lighting) with a clear view of the southern and eastern horizons at least. Bring along a red flashlight (see the Reference Guide) for ease in comparing the planisphere map to your actual view of the night sky.

Set your planisphere for the current month, date, and the time you start observing (remember to express the time as Standard Time). After adapting your eyes to darkness for a few minutes, face South and hold the planisphere above your head, viewing it with the red flashlight.

What constellation or constellations are low in the south at this time? Make a sketch of the southern horizon, including any prominent landmarks (such as lighted towers, church steeples, or smokestacks) along the horizon and any conspicuous star patterns that are visible above it.

> DO NOT just sketch the patterns shown on the PLANISPHERE but stars that you <u>actually see</u> in the night sky, including fainter stars; the planisphere is only a guide for identification.

Given the information from your instructor about the bright planets' locations: **Are any of those planets currently visible?** If so, sketch their locations relative to the horizon and include some of the stars right around them. Also: estimate the direction and altitude of each planet*.

Now turn around, facing North. Again, sketch the view, including the northern horizon and any prominent landmarks, the star *Polaris*, and the locations of the BIG DIPPER and CASSIOPEIA.

> Again: DO NOT just sketch the patterns shown on the PLANISPHERE but stars that you actually see in the night sky, including fainter stars; the planisphere is only a guide for identification.)

The hazy band running across the planisphere's oval section is the *Milky Way*, the band of light which is due to the glow of millions of distant stars in our galaxy. See whether you can glimpse the Milky Way from your location**, and describe what you see.

Exercise 1: Using a Planisphere

Does it have an overall color?
Is its brightness constant all along its length?
Is it stronger in any particular directions or altitudes?

Locate as many of the following stars as you can (some will <u>not</u> be visible when you observe). For each one you **do** find, comment on its location (direction and altitude), any noticeable COLOR, and how noticeably its light FLICKERS (twinkles):

Capella	*Fomalhaut*	*Vega*	*Regulus*
Arcturus	*Spica*	*Sirius*	*Betelgeuse*
Aldebaran			

*   *See the exercise "Altitude and Azimuth."*

**   *This will be difficult from an urban area unless the air is especially clear (such as after a rainstorm) and your location is well away from concentrations of bright lights.*

 See the LAB REPORT CHECKLIST for guidelines in preparing your report.

Exercise 1: Using a Planisphere

Altitude and Azimuth

You Will Need:
- One clear night
- Planisphere*
- Inclinometer (*optional*)**
- Triquetrum (*optional*)***

Purposes: To explore the concept of altitude and azimuth, and to estimate the ALT and AZ for different celestial objects

Background:

The geographical locations of objects on Earth's curved surface may be described by the two quantities latitude and longitude. Latitude is measured perpendicular to Earth's equator (north or south), while longitude is measured parallel to Earth's equator (east or west). Latitude and longitude are really ANGLES originating at Earth's center; see *Figure 1*.

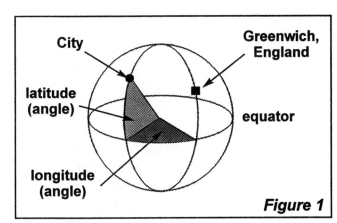

Figure 1

This idea of two perpendicular angles is found again and again in geography, geology, meteorology, astronomy and other physical sciences which deal with curved (spherical) geometry. Just as location on Earth can be defined in terms of latitude and longitude, so can the positions of objects in the sky (day or night) be defined in terms of two perpendicular angles called **altitude** and **azimuth** (*Figure 2*).

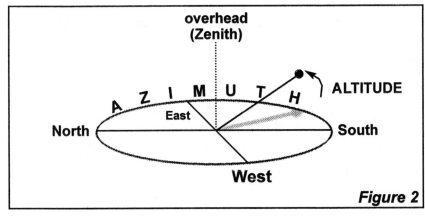

Figure 2

Altitude (ALT) is defined as the angular height of an object above the observer's horizon (assumed to be a flat horizon, of course: no hills or other topography to complcate things). An object with the maximum altitude of 90° is at the observer's

*   See the exercise "Using a Planisphere."
**   See the exercise "Use of an Inclinometer."
***   See the exercise "The Moon's Phases."

Exercise 2: Altitude and Azimuth

overhead point, or zenith, while something with ALT = 0° lies on the horizon and an object with ALT = 45° is located halfway between horizon and zenith.

Azimuth (AZ), as defined in astronomy, is an angle measured ALONG the horizon, starting at North and opening <u>eastward</u>, to the point on the horizon directly BELOW the object. An object above the northern horizon will have an azimuth of 0°, while one above the eastern horizon has AZ = 90° and one with AZ = 135° is above the southeastern horizon.

The object portrayed in *Figure 2* has ALT = 25° and AZ = 150°.
If an object lies <u>at the zenith</u>, its azimuth is <u>undefined</u>. Its altitude is 90°.

Procedure:

Go out on a clear night and locate the BIG DIPPER and the star *Polaris*.

The text below describes a simple way of estimating angles with your fingers. However, if you have done or are doing either the Inclinometer, or the Moon's Phases exercises, you can also measure angles (and with somewhat more accuracy) with the **inclinometer** (for altitude only) or with the **triquetrum** (use horizontally, with one arm pointed toward North, for azimuth; hold vertically, with one arm pointed toward the horizon, for altitude).

The two stars *Merak* and *Dubhe*, in the Bowl of the Big Dipper and farthest from the Handle, are the so-called **Pointer Stars** because a line through them can be extended toward Polaris. However, the Pointer Stars themselves are also useful for estimating angles like altitude and azimuth because the angular distance between them is known.

Refer to *Figure 3*. The angular distance between the Pointer Stars is **5.4.°** Extend

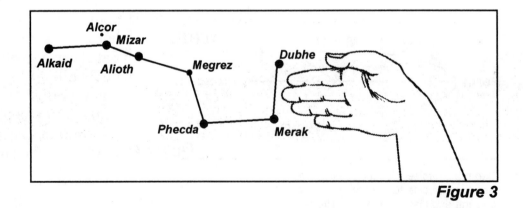

Figure 3

Exercise 2: Altitude and Azimuth

your arm and bend your hand inward so the palm faces you, then rotate your arm and hand until the fingers are pointing perpendicular to the (imaginary) line between the Pointers.

How many fingers just fit between the Pointer Stars? Try to estimate the best fit to "half a finger"; that is, if three are not enough but four extends beyond the Pointers, call the number 3.5. (If four fingers aren't enough, estimate how many more fingers would be needed by spreading one of your fingers.) If the total number of fingers needed to fit between the Pointers is "N," then the angle covered by ONE finger is:

$$5.4° / N$$

To estimate the altitude of an object, hold your hand as shown in *Figure 4* with the object just above your top finger. Shift your hand downward by "four fingers' worth," repeating this until you near the horizon. Keep a count of how many four finger groups there are, including the one from which you started.

When the remaining distance between the bottom edge of your hand and the horizon is less than four fingers' worth, estimate how many fingers' worth remain.

To find the total number of degrees, multiply the total number of fingers used by the number of degrees in ONE finger (formula above).

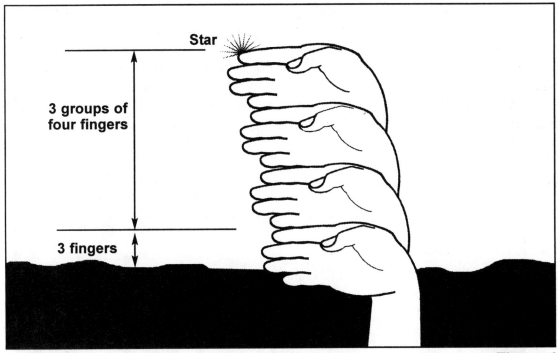

Figure 4

Exercise 2: Altitude and Azimuth

For Example:

You find that between four and five fingers (call it 4.5) is needed to fit between *Merak* and *Dubhe*.

Then the number of degrees represented by ONE finger is:

(5.4° / 4.5) = 1.2°.

The altitude of a rising star is measured by you as 3 full four finger groups plus 3 fingers, or 15 fingers total.

So your estimate of the altitude of the star is:

15 fingers x 1.2° for one finger = 18.°

To estimate azimuth, locate *Polaris*. The point on the horizon directly below it is 0° azimuth. Extend your arm with the hand and fingers vertical, as though you were gesturing for oncoming traffic to stop or pushing against a door. Again, work your way along the horizon "four fingers at a time" until you are just below the object in which you are interested.

Because azimuth has a much wider range (0 to 360°) than altitude, it may be helpful to establish the azimuths of a few local landmarks (such as chimneys or lighted communications towers) visible from your observing location. Then, rather than always re-starting from North, you can start from one of the landmarks (as long as you don't change your observing location). See *Figure 5*. If you have constructed a triquetrum, you can use it to obtain more accurate azimuth angles.

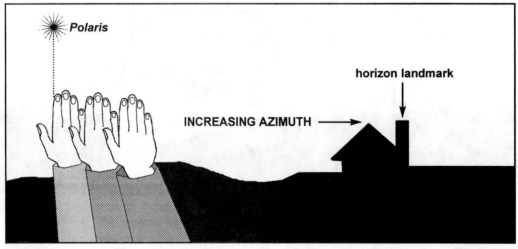

Figure 5

Exercise 2: Altitude and Azimuth

Observations:

Using any of the techniques described above, estimate the ALT and AZ of as many of the following celestial objects as are visible. Some will not be; use your planisphere to help you locate those which are.

It is especially important to record the TIME for each object observed, since the altitudes and azimuths of celestial objects are constantly changing as they rise and set. (If you have established landmarks along your horizon to help you, list them and their azimuths, and provide a sketch of your horizon with the landmarks and their azimuths shown.)

The Moon (if visible)	*Capella*
Aldebaran	*Rigel*
Sirius	*Dubhe*
Regulus	*Arcturus*
Spica	*Vega*
Antares	*Fomalhaut*

  See the LAB REPORT CHECKLIST for guidelines in preparing your report.

Exercise 2: Altitude and Azimuth

Apparent Magnitudes of Stars

You Will Need:
- Planisphere
- Red flashlight
- One clear night (preferably without the Moon visible) for observations

Purpose: To investigate the magnitude system used by astronomers

Background:

When we observe stars (with our eyes alone or with optical aid such as telescopes), we observe their **brightnesses**. A star's brightness is the amount of energy that we receive from it every second over each unit area of our light detector (which could be, for example, the retina of the eye or the mirror of a telescope). But astronomers also use the **magnitude system** when referring to celestial objects, and that is what this exercise explores.

The magnitude system was first developed by the Greek named Hipparcos (or Hipparchus), who lived around 150 B.C. Hipparcos constructed one of the first star catalogs. As part of his work, he divided stars into categories on the basis of their brightness, assigning "first rank" to the brightest stars and "second rank," "third rank," and so on to successively fainter ones. A star's ranking has come to be known as its **apparent magnitude**. Nowadays, the brightest stars have been "promoted" to a ranking of zero.

During the 18th and 19th Centuries, astronomers tried to define the apparent magnitudes of stars more accurately. The British astronomer William Herschel observed stars of different brightnesses with telescopes of different sizes, to determine how much more light one needed to gather from a faint star with a <u>large</u> telescope to make it appear equal in brightness to a bright star viewed through a <u>small</u> telescope; he determined that stars different by FIVE rankings (such as "zero" versus "fifth") had roughly a ONE HUNDRED TO ONE proportion of brightness, with the zero-rank star being the *brighter* one. In 1850, Harvard astronomer Norman Pogson first used a scale in which a difference of five magnitudes was <u>defined</u> as representing an exact 100 to 1 proportion of two stars' brightnesses; this scale has been used ever since.

The magnitude scale is odd. First, because of Hipparchos's ranking system, stars with **low** magnitude numbers (such as 0) are **bright** ones (from which we are receiving a LOT of energy), while those with a **high** magnitude (such as 5) are **dim**; this relationship is backwards! Second (and worse), we speak of magnitude <u>differences</u> between stars (a fifth-magnitude and a zero-magnitude star <u>differ</u> by five magnitudes), but those differences represent one star being so many TIMES brighter than the other (that is, a multiple). If two stars differ by five magnitudes, the brighter one is 100 <u>times</u> brighter. *See Figure 1.*

This star has an apparent magnitude of:

Visibility	5 barely visible to the eye	4	3	2 noticeable but not conspicuous	1	0 one of the brightest stars
Brightness	1 unit	2.5 units	6.3 units	16 units	40 units	100 units

Figure 1

If the magnitude system is this complicated, why do astronomers use it? The main reason is that celestial objects show a huge range in brightness (for example, compare the Sun and Pluto), but the range of <u>magnitude</u> numbers is much smaller; two objects that have a brightness proportion of 10,000 to 1 will only differ by 10 magnitudes.

Observations:

Go out on a clear night to a dark location. Allow your eyes at least five minutes to adapt to the low light level. Using a red flashlight, set your planisphere to the current date and time.

Find a constellation at moderate altitude (say, halfway to the zenith) which contains a few of the brighter stars (larger dots on the planisphere). If the Moon is in the sky, note its phase and location relative to the constellation you chose.

Make a quick sketch of the major star patterns in the constellation, connecting some of them with lines if you wish to mimic the constellation figure on the planisphere (see *Figure 2*, which displays the southern constellation Grus the Crane). Then go back and carefully put in the positions of as many of the fainter stars as you can see (refer to *Figure 3*). At this time, you only need to make smaller marks to indicate the fainter stars.

Now carefully observe the stars you just sketched, this time with the intention of SORTING them into **six** brightness categories. You will do this by **comparing** the brightest stars in the constellation with other bright stars visible in your sky (use your planisphere to help you in locating them).

Stars of magnitude 0 are as bright as the brightest stars visible in your sky*; examples of such stars are *Procyon, Capella, Arcturus* and *Vega.* Some of those stars

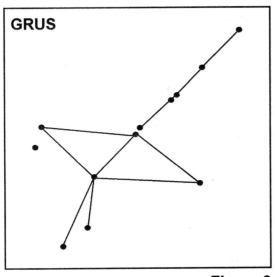 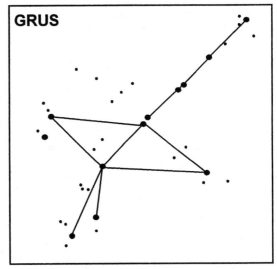

| **Figure 2** | **Figure 3** |

will be visible above your horizon at the time you observe; use the planisphere to help you find them. Are any of the stars in your chosen constellation <u>that</u> bright? If so, call their category "magnitude zero" and use the magnitude range 0 - 5 for your rankings of the other stars in that constellation.

Stars of magnitude 1 are still very noticeable stars, just not as prominent as the stars of zero magnitude. First-magnitude stars include *Aldebaran, Pollux, Regulus, Spica, Antares, Deneb* and *Fomalhaut*; use the planisphere to help you find them. If your brightest stars are in <u>this</u> range, start with "magnitude one" as the brightest category and use the range 1 - 6 for your rankings.

Decide into which of six brightness categories each of the stars you sketched will fit. It may be convenient to use the following descriptions in making your decisions:

Magnitude 0	As bright as the brightest stars visible in the sky *
Magnitude 1	Very noticeable, just not as much so as the very brightest stars
Magnitude 2	Constellation "outline" stars: Not the brightest, but noticeable enough to define the shape of the constellation pattern you sketched. Example stars are: *Polaris*, the three stars in Orion's Belt, *Castor*, and the two "pointer" stars in the Big Dipper.
Magnitude 3	Noticeable, but not very: the "background" stars in the pattern. An example is *Megrez* (the faintest star in the bowl of the Big Dipper).
Magnitude 4	Hard to see in a bright sky, inconspicuous even in a dark sky
Magnitude 5	Barely visible to the eye in a dark sky
Magnitude 6	Good luck seeing these in anything but a really dark sky!

Exercise 3: Apparent Magnitudes of Stars

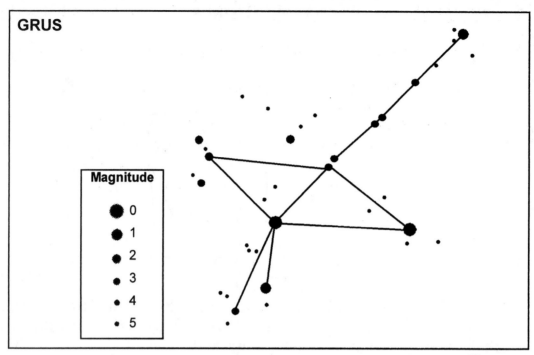

Figure 4

Using different sizes or shapes of symbols to represent the different magnitudes, re-do the star pattern as shown in *Figure 4*.

Questions:

How many stars were there in each magnitude category?

Did you observe any noticeable colors for any of the stars?

In what magnitude category or categories were they found?

 See the LAB REPORT CHECKLIST for guidelines in preparing your report.

---

* The star *Sirius* and the planets Jupiter and Venus are noticeably brighter than magnitude zero; in fact, they are assigned negative magnitude numbers! Ignore these objects in doing this particular exercise.

Exercise 3: Apparent Magnitudes of Stars

The Number of Stars Visible to the Naked Eye *

You Will Need:
- One paper towel tube (or similar tube) about 12" long
- Calculator
- One clear, dark night *and/or* one clear night with bright moonlight

Purposes: To demonstrate how the total number of stars visible to the eye is estimated from samples, and to investigate (if done as a multi-night exercise) how that number changes when the Moon is visible

Background:

For a quick but realistic idea of how many stars a person could see without a telescope, astronomers use a statistical method called *sampling*, in which the numbers of stars in some known FRACTIONS of the sky's area are counted; then that number is "scaled up" to give an estimate of the number that would be visible across the ENTIRE sky. Almost as important as this total number is its uncertainty, which can be calculated from how closely the results from different sampled locations agree with each other.

See *Figure 1*. Imagine looking out at the sky through a hollow tube. The rays of light entering the tube come from a much larger circular spot on the sky. The tube, if moved in all directions, would eventually cover the surface of an (imaginary) hemisphere centered on your head; those directions, extended, would form the "dome of the night sky" above you on any clear night.

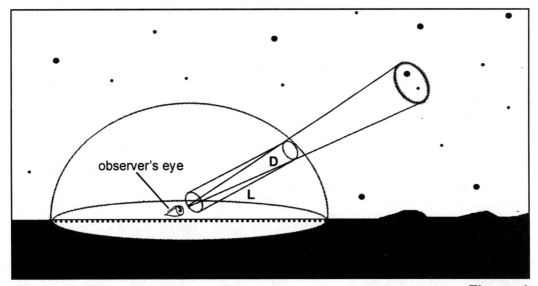

observer's eye

D

L

Figure 1

* *This exercise may be carried out over more than one night.*

Exercise 4: The Number of Stars Visible to the Naked Eye

The length of the tube is L, and its inner diameter is D. The area of the open end of the tube is just the area of a circle of diameter D, which is:

$$\text{Area} = \pi\,(D/2)^2 = \pi\,D^2/4 \quad \text{where } \pi = \text{``pi''} = 3.14159265\ldots$$

If you were to move the tube all around above your head, the far end would trace out the imaginary hemisphere shown in the figure. The radius of this hemisphere would be the length L of the tube. The surface area of an imaginary SPHERE of radius L is:

$$\text{Area of imaginary sphere} = 4\,\pi\,L^2$$

so the surface of an imaginary hemisphere of that radius is:

$$\text{Area of imaginary hemisphere} = 2\,\pi\,L^2$$

The area at the end of the tube, extended and enlarged, projects out onto the **sky**, as shown in the figure. The imaginary hemisphere, extended and enlarged, also projects out as the hemisphere of **sky** seen over an observer at any given time. So we can say that a <u>proportion</u> holds:

$$\frac{\text{Area of hemisphere of sky}}{\text{Area of projected spot}} = \frac{\text{Area of imaginary hemisphere}}{\text{Area of end of tube}}$$

or

$$\frac{\textbf{Area of hemisphere of sky}}{\textbf{Area of projected spot}} = \frac{(2\pi L^2)}{\frac{(\pi D^2)}{4}} = \frac{8L^2}{D^2}$$

Lastly, we make an assumption: that stars are strewn evenly across the night sky. If that is true, then the more area we sample, the more stars we should see. What this means as a formula is that:

$$\frac{\text{Total number of stars in hemisphere of sky}}{\text{Number of stars in projected spot}} = \frac{\text{Area of hemisphere of sky}}{\text{Area of projected spot}} = \frac{8L^2}{D^2}$$

Finally, combining the last two formulas, we have our working formula:

$$\boxed{\begin{array}{l}\textbf{Total number of stars in a}\\ \textbf{hemisphere of sky}\end{array} = \frac{8L^2}{D^2} \times \begin{array}{l}\textbf{(average) number of stars in}\\ \textbf{a projected spot}\end{array}}$$

Exercise 4: The Number of Stars Visible to the Naked Eye

Procedure: Carefully measure L and D for your tube, to better than 1 mm, averaging several measurements, for highest accuracy. Then, calculate the factor $8L^2/D^2$ used in the formula above. For example, if L = 300.5 mm and D = 40.3 mm, then the factor is 8 x (300.5)(300.5) /[(40.3)(40.3)] = 445.)

Observations:

On a clear night (no Moon visible), go to a relatively dark location. Allow your eyes some time to adjust to low light levels - at least 5 minutes, but preferably longer. Then, looking through your tube, count the numbers of stars you can see in each of *at least ten parts of the sky.*

Hints: (1) Look a little out of the side of your eye when counting stars. This technique, called averted vision, lets the eye have a little more sensitivity to faint stars than looking directly at them would (and normally, when you gaze at the stars, your vision isn't restricted by a tube, so some starlight would enter the sides of your eyes).

(2) Try to cover a lot of different directions and altitudes, from looking straight up (zenith direction) to looking near the horizon and toward elevations in-between ; this will give you a more realistic estimate of the star numbers. Carefully **record the direction** (such as NW) **and altitude** (say, 45°) for each observation; an overhead observation should simply be called "zenith."

Suppose you counted stars in ten different areas; lets call the numbers of stars seen in each case N_1, N_2...through N_{10}. Then the **average** number of stars seen through the tube is simply:

$$N_{avg} = (N_1 + N_2 + ... + N_{10}) / 10.$$

This number goes into the working formula on the previous page (N_{avg} = average number of stars in a projected spot).

Calculate a scatter (uncertainty) in your average number; the best choice is σ, the standard deviation. The average number of stars seen through the tube, then, can be expressed as a number with an uncertainty: $N_{avg} \pm \sigma$.

When the average number is multiplied by $8L^2 / D^2$ (in the working formula) to obtain the TOTAL number of stars that should be visible to the naked eye at any one time, *the UNCERTAINTY in the total number is just σ multiplied by the same factor, $8L^2 / D^2$.*

Exercise 4: The Number of Stars Visible to the Naked Eye

> **For Example:**
>
> If Navg = 25 $\pm$ 4 and $8L^2 / D^2$ = 445, then the total number of stars is estimated at 25 (455) $\pm$ 4 (445), or 11,000 $\pm$ 1,800 stars.

Calculate Navg, σ, the estimate of the TOTAL number of stars visible and ITS uncertainty, for the observations you made on the clear, dark night.

<u>To do this exercise as a multi-night project</u>: Repeat your observations (and calculations) on a clear night *when the Moon is visible and bright*. Again, take samples from all around the sky, but take care not to look right near the Moon!

Questions:

[If you did this as a multi-night exercise] How does the total number of stars, and its uncertainty, for the dark (moonless) night compare to those values for the moonlit night?

Can you think of any other reasons that could be responsible for making those numbers change from night to night?

How do you think your results would be affected if you used a tube of different length or diameter to take your samplings?

 See the LAB REPORT CHECKLIST for guidelines in preparing your report.

Use of an Inclinometer

You Will Need:
- Yard stick, meter stick or other long, flat-sided wood
- Protractor
- String or thread
- Small heavy object (such as a nut or fishing sinker) to serve as a plumb weight
- Planisphere (optional)
- One clear night for observations

Purposes: To construct a simple device to measure the altitudes of celestial objects, and to use it to measure the rising motion of a star and obtain an estimate of the local latitude

Background:

An inclinometer is a sighting rod to which is attached an angular scale and some "reference line" which is always vertical, such as a weighted string. Construction of this device is very simple; see *Figure 1*.

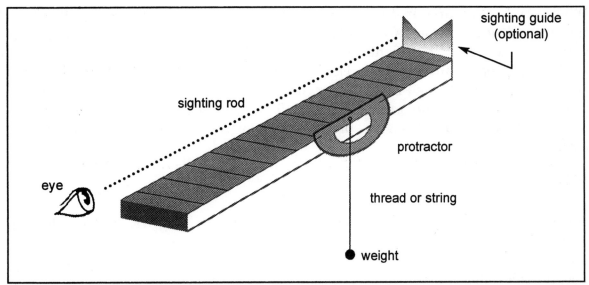

Figure 1

Procedure:

Fasten a thread or string to a protractor, through the small hole in the middle of its flat side (the hole is the pivot point for the angles measured); tie a knot in the string, or fasten the string to the wood with a thumb tack. You can also mount some white paper or cardboard behind the protractor to make reading the angle scale easier.

The "zero-zero" line on the protractor does NOT lie exactly along its flat edge but instead is a bit inside it, at the level of the small hole; those two marks are indicated in *Figure 1*. Attach the protractor to a long, flat-sided piece of wood or similar material *as shown*, so that the two "zero" marks and NOT the protractor's flat edge are level with the edge of the piece of wood. If desired, a sighting guide (see the figure) could also now be attached to the far end of the long sighting rod.

Lay the sighting rod flat at the edge of a level table top and read the angle indicated on the protractor by the hanging string; it will usually be 90 degrees. If it is a bit off from this, your table top may not be level or you may not have mounted the protractor exactly parallel to the sighting rod; correct it if you can. If the angle shown is 90 degrees, then the altitude will be 90° *minus* the number read from the protractor.

For Example:

In *Figure 1*, the sighting rod is shown elevated by about 30 degrees; the weighted thread will, therefore, cross the "60" mark on the protractor scale.

When you are making an actual observation, slowly elevate the sighting rod until the desired object appears at the end of the rod or (if you are using one) at the bottom of the "V" of the sighting notch. Allow a moment or two for the weighted thread to stop swinging, then carefully pinch thread and protractor together, between your thumb and index finger, where the thread crosses the angle scale on the curved edge of the protractor. Carefully turn the protractor and read the number (to the nearest degree, or better if you can) where the thread intersects the scale.

Observations:

On a clear, dark night, find a level location with an unobstructed northern and eastern horizon.

Locate the moderately-bright star *Polaris* in the northern sky (consult your planisphere or ask the instructor for help in finding it). Take several measurements of its altitude and average them.

Record your individual measurement values as well as their average in your lab report.

The altitude of *Polaris* is very nearly equal to the observer's LATITUDE. Consult a national or world atlas in the library to estimate the latitude at the location from which you were observing. **How well did your average agree with that number?**

Exercise 5: Use of an Inclinometer

Look toward the east and find a bright star which is near the horizon. Try to identify this star with the aid of your planisphere. Make a sketch of the horizon, including any prominent landmarks like church steeples, chimneys or smokestacks against which you could compare the star's position at different times.

Record the local time, then make a few observations of the star's altitude. Record the individual measurements as well as the average.

Wait 30 or 40 minutes, then (again recording the time) make a new set of observations of the star's altitude. ALSO take note of any motion of the star <u>left or right</u> relative to the landmarks you sketched.

 Turn in your inclinometer with your lab report. DO NOT dismantle it after doing this experiment; you might use it in other experiments.

Construction and Calibration of a Cross-Staff

You Will Need:
- Construction time
- Meter stick and yardstick
- Plastic pipe or dowel rod
- Scrap lumber or styrofoam
- Pocket calculator with trig functions
- One morning or afternoon for observations

Purpose: To construct a simple device for measuring angles

Background:

Astronomers frequently have to measure the angles separating celestial objects. Some examples are the altitude of a rising or setting star (the number changes with time) and the angular separation of a planet and some bright star (this angle will also change, from night to night, due to the planet's orbital motion).

Perhaps the simplest angle-measuring device, use of which dates back more than 2000 years, is the **cross-staff**. It usually consists of only two parts, though more elaborate versions have also been constructed. The **staff** part is a tube, rod, or stick as long as a person's outstretched arm, and it is usually marked off into inches, centimeters, or other length units. The **crosspiece** is a smaller, movable bar mounted at right angles to the staff and free to move along its length, toward or away from the observer's eye; the ends of the crosspiece may also have sighting notches or projections on them.

Two simple designs for cross-staffs are shown below (*Figure 1*); perhaps you can come up with a better one. The ultimate design, whatever it is, must be fairly

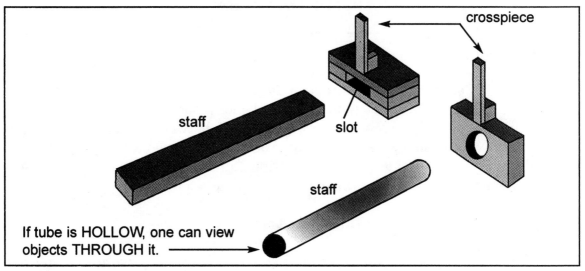

Figure 1

Exercise 6: Construction and Calibration of a Cross-Staff

sturdy; the crosspiece must remain at a right angle to the staff as it is moved.

Procedure:

Design and build a cross-staff where the length of the crosspiece (from center of hole to top) is 100 millimeters (about 4 inches). The staff should have a length of at least one meter (1 meter = 1000 mm). If you use a meter stick, it will already be marked in centimeters and millimeters; otherwise, you will use a ruler to mark it at certain locations (see below) which correspond to particular *angles* subtended by the crosspiece. If you use a tube (such as a length of PVC tubing) for your staff, you can very conveniently look *through* the tube at one of the objects you will be sighting.

Figure 2 shows the basic use of the cross-staff. One end of the staff is held up near the eye, with the crosspiece near the other end. The length L of the crosspiece is a fixed number (100 mm), but the crosspiece can be moved toward or away from the eye; that distance is adjustable, so let's call it X. (X would equal zero if the crosspiece were right up against the eye.)

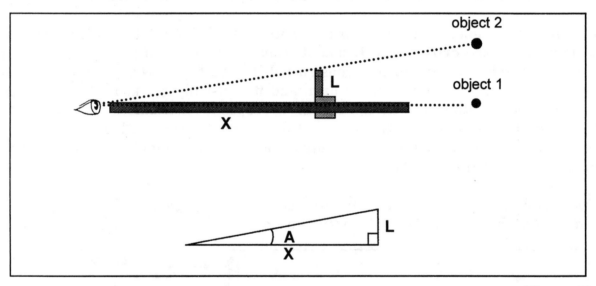

Figure 2

The geometry is shown at the bottom of *Figure 2*. If you are sighting along (or through) the staff at an object (Object 1) and along the end of the crosspiece at another object (Object 2), there is a certain angle A between them. The crosspiece and its length X from the eye make a right triangle, with one side of the right angle equal to L and the other side of the right angle equal to X. The proportion between those two sides is defined as the tangent of the angle we want:

$$\text{tangent (A)} = L / X$$

(If you are not familiar with tangents of angles, a quick way to get familiar is to draw right triangles of the same height but different lengths on graph paper. Let the longest

Exercise 6: Construction and Calibration of a Cross-Staff

direction of the graph paper run left-to-right. Draw a vertical line (say, 100 mm long) which will be the height of every triangle. Now, from the bottom of that line (the origin of your graph), draw a horizontal line toward the right. At various places along the horizontal line (say, at 50 mm, 100 mm, 150 mm and so on) make a mark. For each triangle, draw a diagonal line which connects one of the marks to the top of the vertical line. Use a protractor to measure the angle between the horizontal line and the diagonal line you just drew. The length of the vertical line is L, and the length of the horizontal line from the origin out to the mark is X, so you can find the proportion L/X. Construct a table listing the angles (A) and the values of L/X (which equal the tangent of A), and you will begin to see how the two sets of numbers are related. [For example, when L = X, the proportion L/X equals one and the angle A equals 45°.])

To numerically get the angle from its tangent, we take the arc-tangent* of (L / X).

$$A = \text{arc-tangent} (L / X)$$

For Example:

If L = 100 mm and X = 500 mm, then
A = arc-tangent (100 / 500) = arc-tangent (0.2) = 11.3 degrees

* *On some calculators, arc-tangent is labeled "tan $^{-1}$," while on others one uses the INV (inverse) key together with the TAN key. To test your calculator, try tan $^{-1}$ (1) or INV TAN (1); the result should be 45 (degrees).*

The table below lists the values of X that correspond to different values of A, assuming that L = 100.0 mm.

A (degrees)	X (mm)	A (degrees)	X (mm)
10.0	567.1	16.0	348.7
10.5	539.6	16.5	337.6
11.0	514.5	17.0	327.1
11.5	491.5	17.5	317.1
12.0	470.5	18.0	307.8
12.5	451.1	18.5	298.9
13.0	433.1	19.0	290.4
13.5	416.5	19.5	282.4
14.0	401.1	20.0	274.7
14.5	386.7	20.5	267.5
15.0	373.2	21.0	260.5
15.5	360.6	21.5	253.9

Exercise 6: Construction and Calibration of a Cross-Staff

Using a metric ruler, together with the information in the table, mark your staff *for every 0.5 degree,* starting at the X-value for an angle of 21.5 degrees (this occurs at X = 253.9 mm, a value close to the "near point" for the average eye). You will probably be able to use the cross-staff with the crosspiece extended out to X = 500 or 600 mm (most arms aren't longer!), which means that your range of measurable angles is roughly 10 degrees to 21.5 degrees.

(If you build a cross-staff with a different length for L, you will have to use the tangent formula, specify the values of A (such as 10.0°, 10.5° and so on) and solve for each value of X at which you will mark the staff.)

Observations:

Now use your cross-staff to estimate the height of a building of several stories. Choose a building on level ground and stand right in front of it. Find some feature on the building (such as a windowsill) which is roughly at your eye level.

Move the crosspiece of your cross-staff to about the MIDDLE value in its range. Hold the staff straight out in front of your eye and twist it so that the crosspiece is **vertical.** Sighting along the staff toward the building feature that you estimated as being at eye level, BACK AWAY from the building until the TOP of the building appears roughly level with the TOP END of the crosspiece.

Adjust the crosspiece position (adjust X) until the top of the building is *accurately* (NOT roughly) aligned with the top of the crosspiece. Take several measurements, recording the angle values and their average.

Next, carefully pace off the distance from your observation spot to the building (again, several measurements and an average are advisable). Later, you will need to determine the average length of your stride by pacing off a known distance (such as a hall corridor) that you have also measured with a ruler or tape measure.

Figure 3 shows the geometry of the problem. Your height is h, while the building's height is h + H, where the part H still has to be measured. The distance you stand from the building is D. The triangle formed by L, X and your line of sight to the top of the building is <u>similar</u> to the triangle formed by H, D and your line of sight; when that is true, proportions between the same sides will be equal for both triangles, since both triangles share the same internal angles (here, right angles and the angle A, in particular).

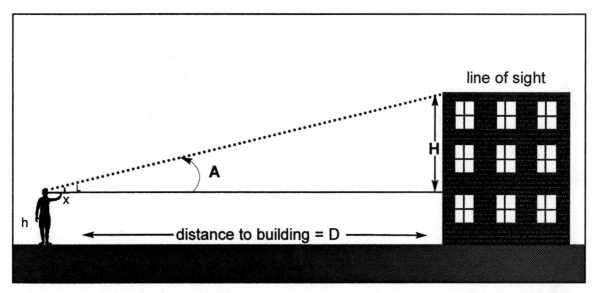

Figure 3

The result is that:

$$L / X = H / D$$

or $$H = (D \times L) / X$$

and | **Estimated height of building** = h + H = h + [(D x L) / X] |

Determine the distance D to the building (in paces, then later in ordinary length units). Measure your own height h (up to your eyes, NOT to the top of your head!) with a ruler or tape measure. (Perhaps a friend can help you with this; if so, mention their contribution in your lab report!). From this information and the above formulas, compute the height of the building.

Question:

Is there some *independent* way of determining the height of this building?

Think about how you could estimate the building's height using simple measurements of building features, like bricks or stair steps. DESCRIBE your approach in the lab report, then OBTAIN that independent estimate of building height. Compare that estimate with the one you obtained from cross-staff measurements; which one do you feel is more accurate, and why?

 TURN IN YOUR CROSS-STAFF with your lab report. Do NOT disassemble it after completing this exercise; it may be useful in other experiments.

Exercise 6: Construction and Calibration of a Cross-Staff

Observing An Eclipse Of The Moon

You Will Need:
- A clear night when a lunar eclipse is expected
- Observing sheets (photocopies of *Figure 3*)
- Binoculars or small telescope
- Red flashlight
- Pencil
- Watch set accurately to the local time

Purposes: To observe changes in the Moon's appearance during its passage into the Earth's shadow, and to time the passage of the Earth's umbral shadow across the Moon's surface

Background:

Every time there is a Full Moon, there is some potential for a lunar eclipse because at each Full Moon, the Sun, Earth, and Moon all lie along nearly the same line (see *Figure 1*). The reason we do NOT have such an eclipse every Full Moon is that the Moon's orbit is tilted slightly to Earth's (about 5°), so that Earth's shadow usually misses striking the Moon. Only when the Full Moon happens to be near or crossing Earth's orbital plane is an eclipse possible.

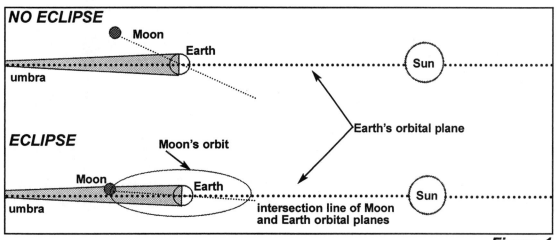

Figure 1

Because the Sun is not a POINT of light but has some physical size, objects like Earth actually cast TWO shadows: a dark, converging cone called the **umbra**, surrounded by a lighter, diverging cone called the **penumbra**. *Figure 2* displays the geometry.

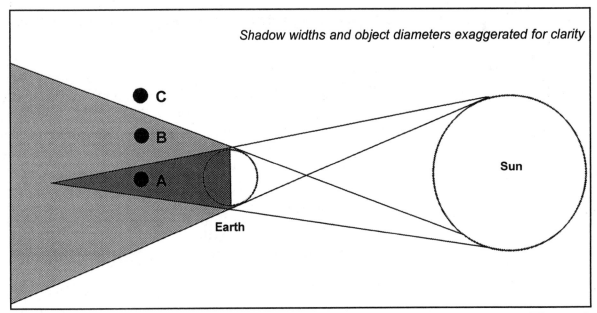

Shadow widths and object diameters exaggerated for clarity

Figure 2

Four different extreme cases of rays of sunlight striking the Earth are shown. Those rays striking the SAME sides of Earth (such as from top of Sun to top of Earth) produce the dark, inner umbra, while the rays which cross to the opposite sides of Earth produce the penumbra.

Someone standing at point A will have all directions to the Sun completely blocked (no light), while someone at point B will have light from the upper parts of the Sun traveling directly to them, so there is less blockage of the light. Finally, someone at point C will have no light blockage.

While the Moon is Full and orbits into point A, there is a **total** *lunar eclipse*. If it happens to orbit into point B, there is what is called a **penumbral** *lunar eclipse*, but most people notice very little change in the Moon's appearance because the amount of dimming in its light is so much less. Sometimes, the Full Moon orbits into a position between A and B, in which case Earth's umbra takes a "bite" out of the Full Moon, and we have a **partial** *lunar eclipse*. The usual situation for the Full Moon is to be out of both shadows (point C), meaning there is no eclipse at all.

If the Full Moon orbits into point A, in principle **it should vanish from sight completely** because Earth is blocking all the sunlight that would otherwise reach it. However, the Moon usually remains visible to us even when totally eclipsed (though sometimes it can get very dark), and the reason is that <u>Earth has an atmosphere</u>. Sunlight (especially the redder light) can seep through Earth's atmosphere into the umbra, giving the totally-eclipsed Moon a reddish glow, which is strongest when the atmosphere is very transparent and largely free of clouds. However, sometimes total lunar eclipses are much darker, with much less light getting through; this occurs whenever Earth's atmosphere has an unusual amount

Exercise 7: Observing An Eclipse Of The Moon

of cloud cover or has been filled with ash particles from a major volcanic eruption.

Procedure:

The table below lists dates, times (EST) of mid-eclipse, and duration of the "bite" (for a partial eclipse) or of totality, for lunar eclipses from 2000 to 2010. Those eclipses in **boldface type** are partly or completely visible from most sections of the continental United States; those primarily visible only from the East or West Coast have an "E" or a "W" designation. Get information from your instructor, the newspaper, or some other source of sky information such as the *Old Farmer's Almanac*, about the details of an upcoming total or partial lunar eclipse.

Date	Partial/Total	EST of Mid-Eclipse	Duration (minutes)	Comments
July 16, 2000	Total	8:57a.m.	106	W: Totality near moonset
January 9, 2001	Total	3.22 p.m.	60	
July 5, 2001	Partial	9:57 a.m.	158	W: Sets in partial eclipse
May 15, 2003	**Total**	**10:41 p.m.**	**52**	
November 8, 2003	**Total**	**8:20 p.m.**	**22**	
May 4, 2004	Total	3.32 p.m.	76	
October 27, 2004	**Total**	**10:05 p.m.**	**80**	
October 17, 2005	**Partial**	**7:04 a.m.**	**56**	W: Most visible before moonset
September 7, 2006	Partial	1:52 p.m.	90	
March 3, 2007	Total	6:22 p.m.	74	E: Rises near totality
August 28, 2007	**Total**	**5:38 a.m.**	**90**	W: Most visible before moonset
February 20, 2008	**Total**	**10:27 p.m.**	**50**	
August 16, 2008	Partial	4:11 p.m.	188	
December 31, 2009	Partial	2:24 p.m.	60	
June 26, 2010	**Partial**	**6:40 a.m.**	**162**	
December 21, 2010	**Total**	**3:18 a.m.**	**72**	

Figure 3 is a Moon map which shows some of the major lunar features, including prominent craters and dark areas or *maria*. **Make about a dozen photocopies of this map.** The map is oriented as you would see the Full Moon if you were looking due south at it; you should be aware that if the Moon is rising when you observe it, its appearance will be rotated somewhat because of its curved path across the sky; you will have to turn the chart counterclockwise a bit to view the features properly; if it is setting, turn the chart clockwise a bit. If you are using a small telescope, you should also be aware that some telescopes will turn the image upside down or flip it right for left.

When you record your data on the observing sheets, be sure to provide specific information about the optical equipment you used: for telescopes, aperture (D) and focal length (F) plus eyepiece focal length (f) and magnification (M = F / f); for binoculars, magnification and aperture (such as "7 X 50.").

Observations:

Look at the Moon frequently through your binoculars or telescope as the eclipse progresses. The Moon enters Earth's umbral shadow from the west, so as the eastern (left) edge of the Moon will be the first portion to show the dark "bite." As the shadow progresses across the Moon's surface, write down the TIME (accurate to the nearest minute) when the shadow covers each of the numbered lunar features completely. Also, make note of any unusual COLOR changes across the Moon's disk at that time (most noticeable for total and deep partial eclipses). **Use a copy of _Figure 3_ for EACH observation of the eclipse you make.**

 See the LAB REPORT CHECKLIST for guidelines in preparing your report.

Lunar Eclipse Observing Sheet

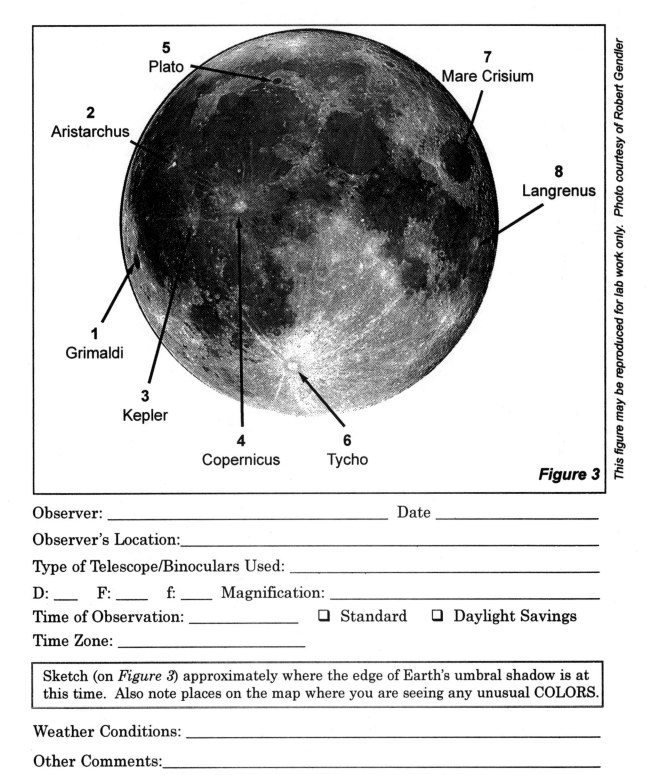

5 Plato
7 Mare Crisium
2 Aristarchus
8 Langrenus
1 Grimaldi
3 Kepler
4 Copernicus
6 Tycho

Figure 3

This figure may be reproduced for lab work only. Photo courtesy of Robert Gendler

Observer: _____ Date _____

Observer's Location: _____

Type of Telescope/Binoculars Used: _____

D: ___   F: ___   f: ___   Magnification: _____

Time of Observation: _____ ❏ Standard ❏ Daylight Savings

Time Zone: _____

Sketch (on *Figure 3*) approximately where the edge of Earth's umbral shadow is at this time. Also note places on the map where you are seeing any unusual COLORS.

Weather Conditions: _____

Other Comments: _____
_____
_____

Exercise 7: Observing An Eclipse Of The Moon

Exploring the Winter Sky

You Will Need:
- One clear night (preferably without a Moon) for observing
- Planisphere
- Binoculars of at least 7x magnification or a small telescope

Purpose: To explore the night sky with binoculars or telescope, using the technique of "star hopping"

Background:

For general viewing of the sky, binoculars and telescopes have some advantages over the eye. Since both have a greater diameter (aperture) than the pupil of the human eye, they have a greater light-gathering power: objects seen through them appear brighter than they would if viewed with the eye alone. One immediate consequence of this is that more stars may be seen as having distinct COLORS, because more of them are brightened above the eye's color vision threshold.

Both binoculars and telescopes gather more light, and also magnify objects' images, to a greater degree than the eye can. However, telescopes generally have such long focal lengths in comparison to binoculars that their magnification is almost TOO large, unless one wants to study planets; the field of view of a telescope is usually restricted to around one degree or less. In contrast, binoculars have reasonable amounts of magnification (say, 7 or 9 times) but a wider field of view (typically, 5 to 10 degrees). Binoculars do not "flip" the images of objects (left for right) as some telescopes do. Also, as BOTH eyes are used when viewing through binoculars, the entire eye-brain sensing/processing system is utilized, leading (some say) to a better contrast between celestial objects and the background sky.

Procedure:

Consult the Reference Guide for information on the magnification and field of view of a pair of binoculars or a small telescope. Though binoculars of 16 to 20x magnifications are manufactured, for general sky-scanning as well as terrestrial viewing it is best to stick with the lower-magnification but lighter weight binoculars, such as 7x35 or 7x50.

Establish the field of view of your binoculars or telescope, in degrees. There are several ways of doing this:

> (1) If the manufacturer specifies the width of the field at a certain distance, you can calculate the field of view (see the Reference Guide).

(2)   Look at the two Pointer Stars (*Merak* and *Dubhe*) in the end of
        the bowl of the Big Dipper (farthest from the stars of the handle).
        Those stars are 5.4 degrees apart. Estimate how many times the
        distance between *Merak* and *Dubhe* will fit across your binocular
        view, then multiply by 5.4 to convert to degrees.

(3)   Look at the Full Moon through your binoculars or telescope. The
        Full Moon's angular diameter is about 0.5 degree. Estimate how
        many Full Moons could fit across your field of view, then
        multiply by 0.5.

(4)   During the day, look at an object (such as a utility pole or house)
        of KNOWN HEIGHT. Move toward it or away from it until the
        object just fills the field of view; then measure your distance from
        the object. If the object's height is "h" and your distance from it is
        "d," the angular field of view is approximately **(h/d) x 57.3
        degrees.**

Figure 1 illustrates the technique of "star hopping" used by amateur astronomers to
point their binoculars or telescopes toward objects which might be too faint to be
easily spotted with the naked eye. A planisphere or star chart is necessary, as is a
knowledge of the angular field of view of the binoculars or telescope. The trick to star
hopping is to start with a bright star, then move it to an edge of the field of view so
that another star becomes visible near the opposite edge; it in turn becomes the new
starting point for hopping to still other stars.

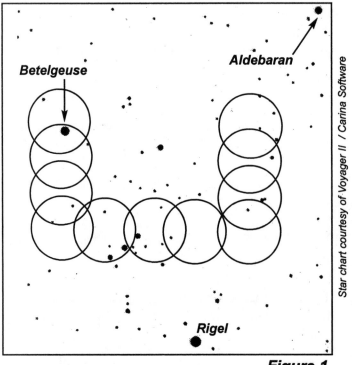

Figure 1

Exercise 8: Exploring the Winter Sky

In the figure, you are looking at objects in the constellation Orion. Starting with the bright star *Betelgeuse*, you proceed downward toward the region of Orion's Belt; then you turn toward the right (westward) and eventually upward again as you explore the fainter stars in Orion's Shield.

It is useful to always move your binoculars around in a north-south and east-west fashion because this duplicates the directions used on star charts. However, you should be aware that on star charts, "north-south" refers to the direction between the celestial poles and "east-west" to the path followed by rising and setting objects. At our (temperate) latitude, the "north-south" direction is ONLY vertical for objects lying due north or due south in the sky; if a constellation is observed while RISING, the "north-south" direction is actually TILTED toward the left. (*See Figure 2.*)

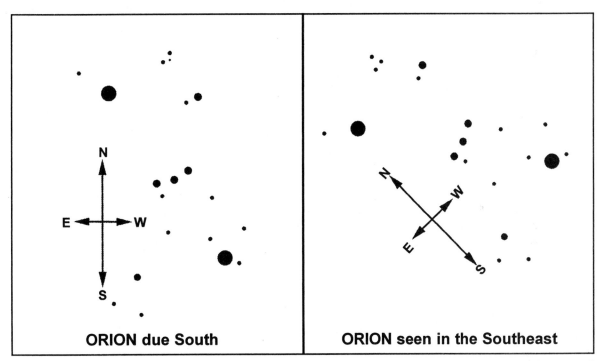

ORION due South   **ORION seen in the Southeast**

Figure 2

Observations:

Figure 3 shows the southern sky as it would be seen from mid-Northern latitudes on evenings of late fall or early winter: specifically, around midnight (Standard Time) in mid-December, 10 p.m. in mid-January and 8 p.m. in mid-February. North is at the top, East to the left.

Establish the field of view of your binoculars or telescope, in degrees.

Use the planisphere to help you identify the constellations and bright stars in this part of the sky. Starting from one of the bright stars, "star hop" to each of the

Exercise 8: Exploring the Winter Sky

CIRCLED AREAS on the chart and observe the object or objects there.

For each circled area:

- Make a sketch of what you see there, in the circular field of view of your binoculars or telescope.

- Describe the appearance of the object(s), such as color or shape.

- Note the date, time, and weather conditions.

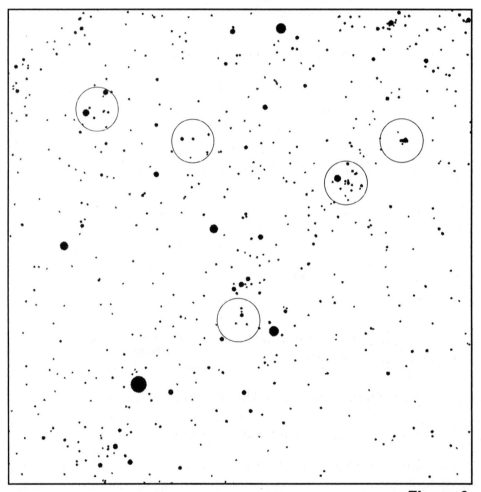

Figure 3

 See the LAB REPORT CHECKLIST for guidelines in preparing your lab report.

Exercise 8: Exploring the Winter Sky

Exploring the Spring Sky

You Will Need:
- One clear night (preferably without a Moon) for observing
- Planisphere
- Binoculars of at least 7x magnification or a small telescope

Purpose: To explore the night sky with binoculars or telescope, using the technique of "star hopping"

Background:

See the exercise "Exploring the Winter Sky" for information on star hopping and finding the field of view of your equipment.

Establish the field of view of your binoculars or telescope, in degrees.

Observations:

Figure 1 shows the southern sky as it would be seen from mid-Northern latitudes on evenings of late winter or early spring: specifically, around midnight (Standard Time) in mid-March, 10 p.m. in mid-April and 8 p.m. in mid-May. North is at the top, East to the left.

Use the planisphere to help you identify the constellations and bright stars in this part of the sky. Starting from one of the bright stars, "star hop" to each of the CIRCLED AREAS on the chart and observe the object or objects there.

For each circled area:

- Make a sketch of what you see there, in the circular field of view of your binoculars or telescope.

- Describe the appearance of the object(s), such as color or shape.

- Note the date, time, and weather conditions.

 See the LAB REPORT CHECKLIST for guidelines in preparing your lab report.

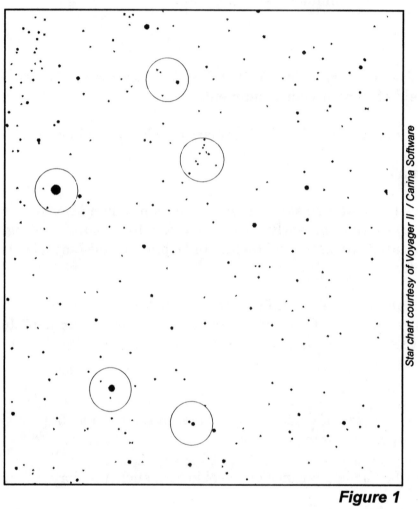

Star chart courtesy of Voyager II / Carina Software

Figure 1

Exploring the Summer Sky

You Will Need:
- One clear night (preferably without a Moon) for observing
- Planisphere
- Binoculars of at least 7x magnification or a small telescope

Purpose: To explore the night sky with binoculars or telescope, using the technique of "star hopping"

Background:

See the exercise "Exploring the Winter Sky" for information on star hopping and finding the field of view of your equipment.

Establish the field of view of your binoculars or telescope, in degrees

Observations:

Figure 1 shows the southern sky as it would be seen from mid-Northern latitudes on evenings of late spring or early summer: specifically, around midnight (Standard Time) in mid-July, 10 p.m. in mid-August and 8 p.m. in mid-September. North is at the top, East to the left.

Use the planisphere to help you identify the constellations and bright stars in this part of the sky. Starting from one of the bright stars, "star hop" to each of the CIRCLED AREAS or the OBJECTS INDICATED BY ARROWS on the chart and observe the object or objects there.

For each circled area:

- Make a sketch of what you see there, in the circular field of view of your binoculars or telescope.

- Describe the appearance of the object(s), such as color or shape.

- Note the date, time, and weather conditions.

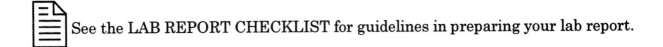 See the LAB REPORT CHECKLIST for guidelines in preparing your lab report.

Exercise 10: Exploring the Summer Sky

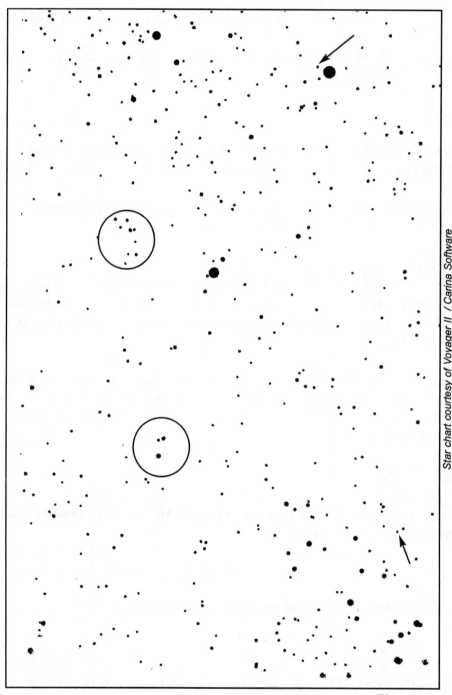

Star chart courtesy of Voyager II / Carina Software

Figure 1

Exercise 10: Exploring the Summer Sky

Exploring the Autumn Sky

You Will Need:
- One clear night (preferably without a Moon) for observing
- Planisphere
- Binoculars of at least 7x magnification or a small telescope

Purpose: To explore the night sky with binoculars or telescope, using the technique of "star hopping"

Background:

See the exercise "Exploring the Winter Sky" for information on star hopping and finding the field of view of your equipment.

Establish the field of view of your binoculars or telescope, in degrees

Observations:

Figure 1 shows the southern sky as it would be seen from mid-Northern latitudes on evenings of late summer or early autumn: specifically, around midnight (Standard Time) in mid-September, 10 p.m. in mid-October and 8 p.m. in mid-November. North is at the top, East to the left.

Use the planisphere to help you identify the constellations and bright stars in this part of the sky. Starting from one of the bright stars, "star hop" to each of the CIRCLED AREAS or the OBJECTS INDICATED BY ARROWS on the chart and observe the object or objects there.

For each circled area:

- Make a sketch of what you see there, in the circular field of view of your binoculars or telescope.

- Describe the appearance of the object(s), such as color or shape.

- Note the date, time, and weather conditions.

 See the LAB REPORT CHECKLIST for guidelines in preparing your lab report.

Exercise 11: Exploring the Autumn Sky

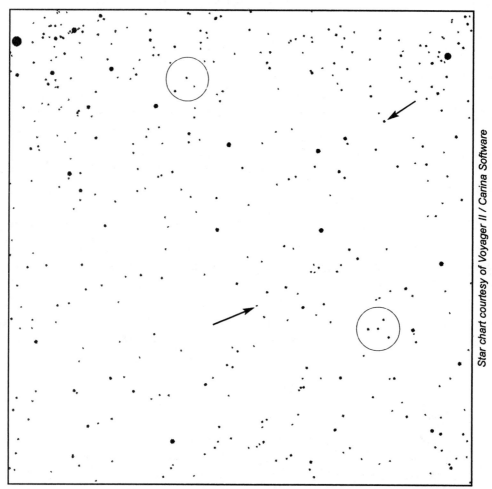

-72-

Star chart courtesy of Voyager II / Carina Software

Figure 1

Exercise 11: Exploring the Autumn Sky

Observing Double Stars

You Will Need:
- Telescope, or binoculars of at least 7x magnification
- Tripod or other support
- Star charts
- Planisphere
- One or two clear nights for observing

Purpose: To observe the brightnesses, colors and other properties of star pairs

Background:

Our Sun is somewhat unusual in being a single, isolated star; more often, in our galaxy, stars form or are found in <u>groups</u>. Concentrated groupings of stars in galaxies are called *star clusters* ; two good examples are the *Pleiades* or Seven Sisters cluster, and (near it in the early winter sky) the *Hyades* cluster (the head of Taurus the Bull). These clusters can contain hundreds of member stars, separated by (typically) a few light-years.

Groups of stars with fewer members and with much smaller separations (tens to thousands of Astronomical Units*) are called *multiple star systems.* Our nearest neighbor [unfortunately not visible from mid-Northern latitudes] is the **Alpha Centauri system**, which is about 4.3 light-years distant. It is a three-star system containing a beautiful pair of golden-yellow stars (Alpha Centauri A and B) separated (on average) by 23 A.U.; orbiting this pair at a much greater distance (about 10,000 A.U.) is a third object, a dim red star called Proxima Centauri.

A system which only contains <u>two</u> stars is called a *binary system.* Sometimes the stars orbit very close to each other (separations of only a few A.U.), and sometimes they have much wider orbits (hundreds to thousands of A.U.). Whatever the reasons for such differences, the binary systems just mentioned share one important characteristic: the stars in them are *bound to each other by their mutual gravitation* and orbit their *common center of mass;* see *Figure 1,* where the stars' relative orbital speeds are shown with arrows.

---

* One Astronomical Unit (A.U.) is the average distance between Earth and Sun: about 93 million miles or 150 million kilometers. One light-year equals 63,200 A.U. The Alpha Centauri star system lies about 274,000 A.U. from our Sun. { The A.U. is a useful unit for separations of stars in multiple star systems, while the light-year is generally better suited to distances between stars in star clusters, or between unconnected star groups (such as the Sun and the Alpha Centauri system) in our galaxy. }.

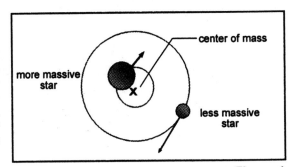

Figure 1

Exercise 12: Observing Double Stars

The term "double star" is a generic one, referring to an observation one could make of two stars which seem close together when viewed through binoculars or a telescope. Sometimes, such pairs of stars ARE really close together, which makes them a binary system (see above). However, there are also cases where one of the two stars is much more distant than the other, and the observer views a chance alignment of two unrelated and unconnected stars. Such double stars are known as optical doubles and are not of much interest to astronomers other than looking pretty; on the other hand, the study of binary systems gives astronomers information about the masses of stars (sometimes also their sizes) and the workings of gravitation outside our solar system.

How can one distinguish a true binary system from an optical double? If the pair is a binary system and the stars complete their orbits in less then (say) 100 years, the effects of orbital motion may become noticeable after five or ten years. The orientation of one star relative to the other, and the observed separation of the two stars, will change; see *Figure 2*. { Such changes will not be seen for optical doubles.}

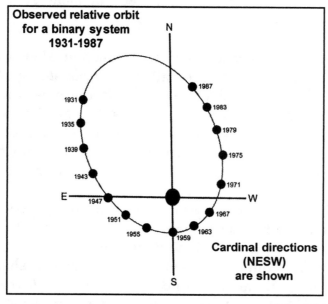

Figure 2

Binary systems with larger orbits tend to take more time to complete them. Some binaries may require 100,000 years or more! It is virtually impossible to observe the position changes of such stars due to their orbital motion, even over a full human lifetime. Fortunately, the stars' *common proper motions* (slow changes in position on the sky due to the system's motion through space) allow astronomers (after several decades!) to state that they are a connected pair. { The proper motions of stars in optical doubles usually show no great similarity, either in direction or amount; see *Figure 3*. }

Exercise 12: Observing Double Stars

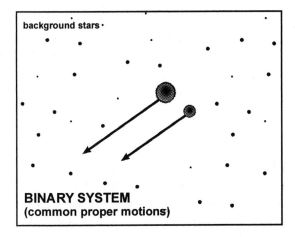

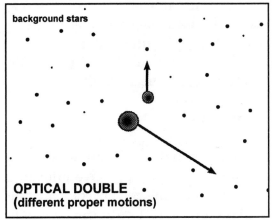

BINARY SYSTEM
(common proper motions)

OPTICAL DOUBLE
(different proper motions)

Figure 3

Procedure:

There are many double stars within the reach of binoculars or small telescopes.

The eye alone, without any other optical aid, can "split" (see as distinct points of light) the stars *Mizar* and *Alcor** in the handle of the Big Dipper; they are only separated by about 0.2 degree. With binoculars of 7x magnification (if held steadily or supported), one can distinguish stars seven times closer together. With a telescope used at 40x magnification, one can distinguish stars about forty times closer together.

For angles less than 1 degree, astronomers use subdivisions called **arc-minutes** and **arc-seconds**. One degree equals 60 arc-minutes, and one arc-minute equals 60 arc-seconds; there are, then, also 3600 arc-seconds in one degree. In symbolic form,

$$1^0 = 60\ ' = 3600\ "$$
$$\text{and} \qquad 1\ ' = 60\ "$$

where ^{0}, $'$, $"$ are the symbols for degrees, arc-minutes and arc-seconds, respectively. In this notation, the different optical devices mentioned above can split double stars separated by the following angles**:

EYE: can split angles greater than 12 ' (= 720 ")
7x BINOCULARS: angles greater than 1.7 ' (= 103 ")
40x TELESCOPE: angles greater than 0.3 ' (= 18 ")

The stars in the list which follows were all chosen to be easily split in binoculars or a small telescope. The columns in the table list the *double star name*, the *constellation* in which it is found, the *magnitude**** of each star of the pair, the

Exercise 12: Observing Double Stars

separation (in arc-seconds) of the two stars, and (when available) *remarks* on whether the pair has common proper motions or has been identified as an optical double. A letter "T" under Remarks means that a telescope MUST be used to observe the double; Gamma Delphini (*see table*) will require higher magnification than the others to be split successfully.

Observations:

Pick at least half a dozen double stars from the list and observe them with binoculars or a telescope ; give information about the optical equipment used in your lab report, including the size of your field of view. (See the Reference Guide.) Consult your planisphere to locate the constellations on the particular dates and times you observe, then use the charts of those constellations (*Figures 4 - 7*) to locate the double stars themselves.

* See Figure 3 in the exercise "Altitude and Azimuth" for an identification chart.
** These are approximate numbers; experienced observers can do somewhat better.
*** See the exercise "Apparent Magnitudes of Stars."

It is **essential** when doing this exercise that your optical equipment have a sturdy support, such as a tripod mount. A jittery telescope or binoculars will prevent you from seeing the stars as distinct pairs. (Binocular supports are commercially available; at the very least, you should steady them against a wall or tree once you have located an object.)

Make a drawing of each double star field that you observe. The field of view will be round, and the double star will not take up much of it. Sketch the double star and also any other conspicuous stars in view.

Questions:

How bright is one star compared to the other?

Does either star (or do both) show noticeable COLOR?

Were the stars easy or hard to split?

How was the double star ORIENTED in the field? (Imagine a line through the pair of stars. How was that that line tilted relative to the top of your field of view?)

What direction were you facing when making the observation? How high in the sky (altitude) was the double star?

 See the LAB REPORT CHECKLIST for guidelines in preparing your report.

Exercise 12: Observing Double Stars

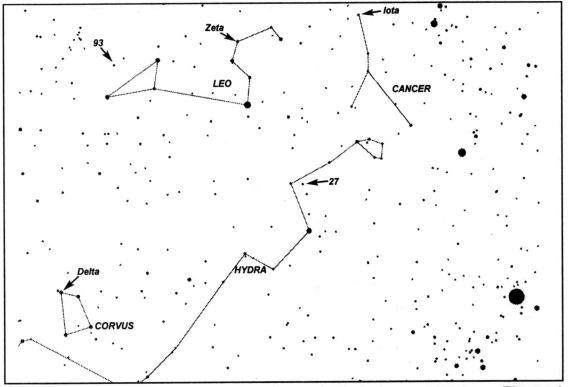

Star charts courtesy of Voyager II / Carina Software

Figure 4

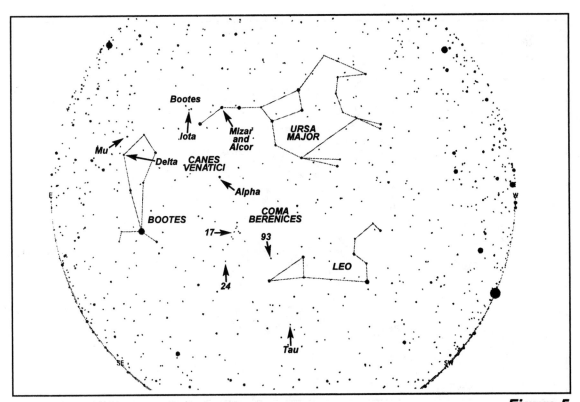

Figure 5

Exercise 12: Observing Double Stars

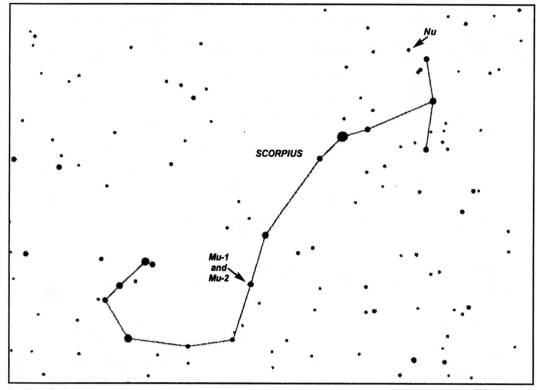

Figure 6

Star charts courtesy of Voyager II / Carina Software

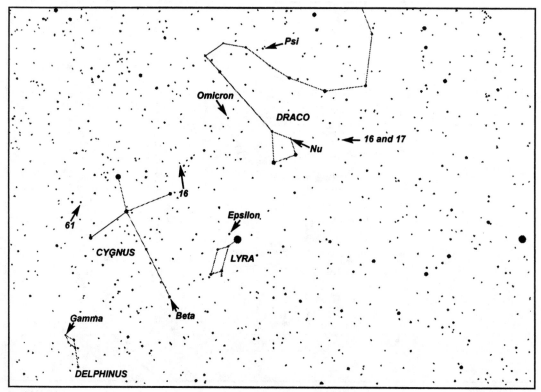

Figure 7

Exercise 12: Observing Double Stars

Double Stars for Binoculars and Telescopes

Double Star	Constellation	Magnitudes		Separation	Remarks
Iota (ι) Cancri	Cancer	4.2	6.6	31"	T
27 Hydrae	Hydra	5.0	6.9	232	c.p.m. pair
Zeta (ζ) and 39 Leonis	Leo	3.4	6.0	326	optical double
Tau (τ) Leonis	Leo	5.0	7.6	91	
93 Leonis	Leo	4.6	8.9	74	T
17 Comae	Coma Berenices	5.3	6.7	145	c.p.m. pair
Delta (δ) Corvi	Corvus	3.0	9.2	24	c.p.m. pair; T
24 Comae	Coma Berenices	5.0	6.6	20	c.p.m. pair; T
Alpha (α) Canum Venaticorum	Canes Venatici	2.9	5.5	19	c.p.m. pair; T
Mizar and Alcor	Ursa Major	2.1	4.0	708	c.p.m. system
Iota (ι) Bootis	Bootes	4.8	7.5	39	T
Delta (δ) Bootis	Bootes	3.5	7.9	105	c.p.m. pair
Mu (μ) Bootis	Bootes	4.3	6.5	109	c.p.m. pair
16 & 17 Draconis	Draco	5.5	5.6	90	c.p.m. pair
Mu-1 (μ¹) and Mu-2 (μ²) Scorpii	Scorpius	3.1	3.6	346	c.p.m. pair
Nu (ν) Scorpii	Scorpius	4.2	6.1	41	quadruple system *
Nu (ν) Draconis	Draco	4.9	4.9	62	c.p.m. pair; T
Psi (ψ) Draconis	Draco	4.6	5.8	30	c.p.m. pair; T
Epsilon (ε) Lyrae	Lyra	4.7	4.6	208	quadruple system *
Omicron (o) Draconis	Draco	4.7	8.1	35	optical douple; T
Beta (β) Cygni	Cygnus	3.1	5.1	34	c.p.m. pair; T
16 Cygni	Cygnus	6.0	6.2	39	c.p.m. pair; T
Gamma (γ) Delphini	Delphinus	4.5	5.5	10	c.p.m. pair; T
61 Cygni	Cygnus	5.2	6.0	30	c.p.m. pair; T

* The two close pairs can be split in large telescopes

Exercise 12: Observing Double Stars

Astrophotography: Star Colors

You Will Need:
- 35mm camera with time-exposure capability
- Shutter cable
- Tripod or other support
- Color slide or print film
- Planisphere
- One clear night (preferably without a Moon) for observing

Purpose: To explore the relation between the colors of stars as recorded on film and their spectral types

Background:

From the middle of the 19th Century onwards, astronomers began to analyze the chemical elements in the atmosphere of stars, adapting techniques first used for element identification in chemical laboratories (see also the exercise "Spectra With a Grating or Compact Disk").

The spectrum of a star is usually a *dark-line spectrum*: a continuous "rainbow" of colors with a pattern of dark lines or gaps overlaid. In the last decades of the 19th Century and the first of the 20th, much work was done at Harvard (then Harvard College) to organize and classify the patterns of lines. Essentially, the patterns were assigned letters in the alphabet ("spectral types") based on the **simplicity** of the pattern: Types A and B showed simple line patterns, while Type G was more complex, and Type M showed a very complex pattern (*see Figure 1*).

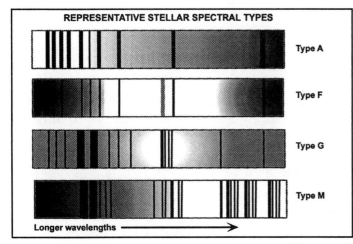

Figure 1

Around 1920, physicist Meghnad Saha showed that the differences in patterns of dark lines from one star to another were mainly due to differences in **surface temperature**, which has an effect on the ability of atoms of different chemical elements to absorb visible light. For example, most of the dark lines in stars of Type A are due to hydrogen, which absorbs light well at temperatures around 10,000° Kelvin (about 17,000° Fahrenheit); on the other hand, stars of Type G have many lines due to iron and other metals, which absorb light best around 6,000° Kelvin (around 10,000° Fahrenheit). Once the role of temperature in affecting line patterns was understood,

Exercise 13: Astrophotography: Star Colors

the spectral classification sequence (A, B, ...) was rearranged *in order of decreasing surface temperature*, into the form now used by astronomers: **O, B, A, F, G, K, M**.

Stars of Type O have the highest surface temperatures: four to five times that of our Sun. Stars of Type B are not as hot, but are still hotter than Type A. Stars of Type A are about twice as hot as the Sun. *The Sun is a Type G star.* The coolest stars are of Type M, with roughly half the Sun's surface temperature.

In the first decade of the 20th Century, astronomer Henry Norris Russell devised a mnemonic (jingle) for remembering the modern sequence of spectral types. It is:

> ## Oh Be A Fine Girl (or Guy), Kiss Me.
> *(You can certainly create your own versions of this!)*

Because so many spectra are available for stars, it is now possible to classify their line patterns even more accurately than just seven categories, so astronomers have divided each spectral type into ten sub-types, represented by a number (from 0 to 9) following the spectral type letter, such as G2. Stars with a lower number are at the hotter end of their spectral type; for example, a star of Type G2 is hotter than a star of Type G8, even though both are of Type G ("Sun-like") overall. It is interesting to realize that spectral classification for stars is accurate enough for us to estimate the temperature of almost any star we can see to within about 300° (Kelvin)!

Shortly after Saha's work resulted in the rearrangement of the spectral type sequence to OBAFGKM, it also became clear to astronomers that the rearranged sequence was ALSO closely related to the overall observed COLORS of stars. You will be investigating that relationship in this exercise.

Observations:

You will photograph one of three constellations: Ursa Major (any season), Cygnus (summer and autumn), or Leo (winter and spring). Use your planisphere to help you choose which one is most easily visible on the date and time you choose to observe it.

Load your camera with film of moderate speed (ASA 100 or 200). If you use <u>color slide</u> film, the star colors are generally truer because you don't have to go through the intermediate step of printing. On the other hand, *print* film is somewhat more readily available, and prints are easy to examine. The type of film you use is your choice. (Note: in either case, you may have to give the photo processing lab special instructions about cutting, mounting, or printing your exposures.)

Choose a night and location which are fairly dark, so sky brightness (which tends to

Exercise 13: Astrophotography: Star Colors

reduce contrast on astrophotos) is minimized. Mount your camera on a tripod, set the shutter for time exposures, attach a shutter cable, and open your lens to its widest aperture. You will want to take several exposures of your chosen constellation - say, 10, 20, 30 seconds, and 1, 2 minutes - to record some of the fainter stars. Because your camera is on a fixed mount, Earth's rotation will cause the stars' images to TRAIL a little (become elongated) across the photograph; this will not affect your ability to do the exercise.

Normally, for astrophotography, the focus of your camera lens should be set at "infinity." However, to better record the COLORS of stars, you might want to try making the stars' images larger by putting them *slightly out of focus.* You will lose some of the fainter stars, but the colors of brighter stars will be more faithfully recorded.

Figures 2, 3, and *4* show the major stars in Ursa Major, Cygnus, and Leo, with spectral types marked for each. After you have gotten your photographs processed, choose the best photograph (the one which shows the most star **colors**) and try to match the stars with those on the appropriate chart.

Make a list of spectral types and colors for the stars you identified in your constellation, and also indicate for each star whether it is <u>bright</u>, <u>moderate</u>, or <u>faint</u> on the photograph.

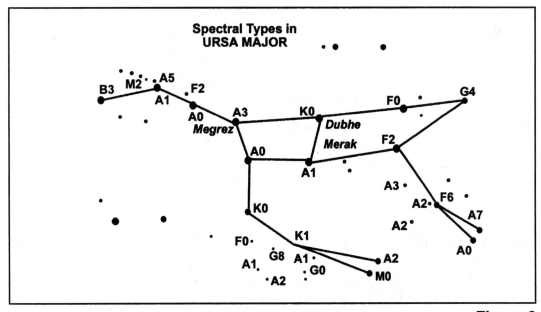

Figure 2

Exercise 13: Astrophotography: Star Colors

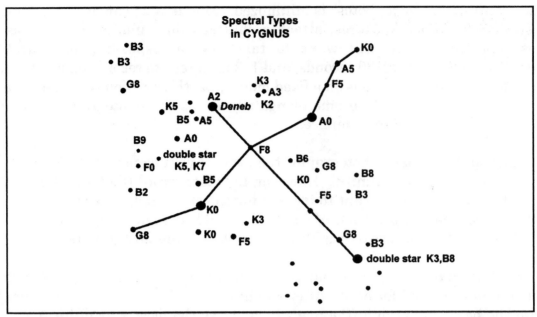

Figure 3

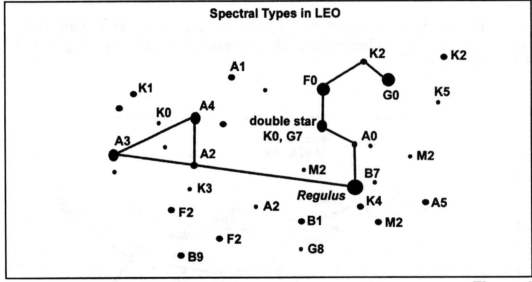

Figure 4

Question:

How does the spectral type of a star relate to its color? Answer for all the stars.

Now, *EXCLUDE* the stars you categorized as "faint." Is there a <u>better</u> relationship between color and spectral type for the stars that remain? There might be; colors of fainter stars are sometimes not recorded as faithfully by color film.

 See the LAB REPORT CHECKLIST for guidelines in preparing your lab report.

Exercise 13: Astrophotography: Star Colors

Measuring The Moon's Orbital Motion

You Will Need:
- Calibrated cross-staff*
- Calculator
- Planisphere (optional)
- Brief observations on two successive clear nights

Purposes:
To estimate the Moon's position with respect to a bright star, and from changes in that position over two or three nights to estimate the Moon's orbital period

Background:

The Moon's orbit around the Earth forms the basis for the lunar calendar of many cultures and, loosely, for our calendar month (or "moonth"). But the month is based upon the time for one complete cycle of lunar phases, say from Full Moon to the next Full Moon (one **lunation**, which takes 29.53 days), rather than upon the actual time required for the Moon to complete one 360-degree orbit around Earth.

The Moon's orbital period is NOT the same length as one lunation because of the motion Earth and Moon also have around the Sun; both Earth's orbital motion around the Sun and the Moon's orbital motion around Earth affect the Moon's phases.

To measure ONLY the Moon's orbital motion requires use of some very distant reference point, one which is not connected to Earth, Sun or Moon at all. A bright star is such a suitable reference point.

Procedure:

Refer to *Figure 1* on the following page. Consult your local newspaper for the date of the next New Moon and the date of the First Quarter Moon that follows it. **The most useful time to attempt observations will be from 2-3 days after New until a few days past First Quarter.** If you wait much longer than that, the Moon may rise very late in the evening -- still observable, of course, but you'll have to stay up later! The other problem is that the Moon is very bright near Full phase, lighting up the night sky around it, and it may be harder to find a suitable bright star near it.

---

* See the exercise "Construction and Calibration of a Cross-Staff."

Figure 1

Observations:

On your first night of observing, look west or southwest just after sunset to find the waxing crescent Moon. Try to find a fairly bright star not far from it, *to its left and slightly higher*. (Note: You will need to identify this same star the next time you observe! A planisphere may help.)

Aim the staff of your cross-staff directly toward the center of the Moon's disk, and turn the device so that the top of the crosspiece is aimed toward the bright star. Adjust the crosspiece until the bright star lies exactly at the end of the crosspiece while the staff points toward the Moon's center.

Take several measurements of the angle and average them. Be sure to record the TIMES of your observations, as well as the date, weather conditions, and any other useful information.

On the following night, if possible, go out at around the SAME TIME and locate the Moon and the bright star used on the previous night. (You will also want to record how much the Moon's phase has progressed in appearance.) Now the bright star will probably be to the RIGHT of the Moon (see *Figure 1*), so you will need to face the crosspiece toward the right when making a measurement. Again, as on the first night, measure the angle between the Moon and the bright star, **taking several readings and averaging them**. List ALL measurements, not just their average, in your lab report! Also, keep careful note of the TIME!

If it is not possible to observe the Moon on two consecutive nights, you may try to observe it on a third night; but those observations will be more difficult because the Moon will likely have moved out of the angle range of your cross-staff. It's really better to keep looking for two consecutive nights, even if you can't always observe at

Add together the average angle measures for the two nights and call the result angle A:

A = average angle (for night 1) + average angle (for night 2)

Compute the elapsed time (in hours and minutes) between your first and second nights' observations, in hours and minutes, and call that result T:

T = time elapsed between the two nights' observations

Convert T into DAYS, using the fact that 1 day = 24 hours = 1440 minutes

---

For Example:
If T = 22 hours 56 minutes, then T also equals (22/24) day + (56/1440) day; or **T = 0.956 day**

To find the Moon's orbital period in days, use the proportion

$$\frac{A}{T} = \frac{360°}{\text{Moon's orbital period}}$$

or **Moon's orbital period (days) = 360° x $\frac{T}{A}$**

Compare your results with the accepted value of 27.32 days, as a percentage difference:

Percentage difference** = { $\frac{[\text{ your estimate - 27.32 days }] \times 100}{27.32 \text{ days}}$ } percent

Questions:

How close (%) did you come?

What do you feel were the greatest sources of uncertainty?

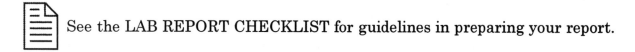 See the LAB REPORT CHECKLIST for guidelines in preparing your report.

** *Negative differences are allowed, and SHOULD BE reported as negative, if your estimate is lower than the published value.*

Exercise 14: *Measuring The Moon's Orbital Motion*

Spectra with a Grating or Compact Disk

You Will Need: • Slide-mounted transmission grating (commercially available) or compact disk (music, CD-ROM, etc.)
    • One or two evenings for observations

Purpose: To investigate the types of spectra of a variety of light sources

Background:

Isaac Newton was among the first people to appreciate and make use of the fact that "white light" (a mixture of all colors), when passed through glass, water or other clear, dense substances, was *dispersed* into its separate colors; that is, a **spectrum** was produced. (The plural form of the word "spectrum" is "spectra.") Light can be considered as energy spreading in waves from a source (such as a star), and the distance between each wave, called the **wavelength**, determines what color the light has. Longer waves mean redder light.

The use of glass prisms to examine the spectra of stars was first applied in the mid-1800s, by people like Angelo Secchi, while around the same time in world laboratories chemical elements were being identified in terrestrial substances from their spectra.

Gustav Kirchhoff, who first identified the element gold in the spectrum of the Sun in 1860, found that hot, luminous objects tended to show one of three kinds of spectrum: (1) a **continuous spectrum** (all wavelengths of light represented), (2) a **dark-line** or **absorption spectrum** (a continuous spectrum with dark "gaps" at certain wavelengths), or (3) a **bright-line** or **emission spectrum** (all of the light radiated at only a few different wavelengths). These spectra are shown in *Figure 1*.

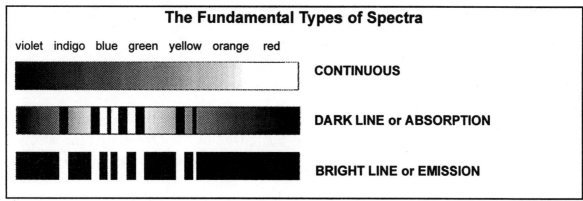

The Fundamental Types of Spectra

violet indigo blue green yellow orange red

CONTINUOUS

DARK LINE or ABSORPTION

BRIGHT LINE or EMISSION

Figure 1

A *continuous spectrum* originates in hot, dense objects, such as metal light-bulb filaments, volcanic lava, or the deeper parts of stars where the gases are

Exercise 15: Spectra with a Grating or Compact Disk

enormously compressed.

A *dark-line spectrum* is caused by thinner (and usually cooler) gases surrounding one of the first groups of objects, such as an atmosphere surrounding a star. In such an atmosphere, individual atoms are far enough apart to be able to absorb particular wavelengths of light to reorganize their internal structure (move their electrons into different energy levels) before colliding with other atoms.

If the star's atmosphere could be viewed BY ITSELF (against a black background rather than in front of the star), one could see those particular energies being RELEASED again from the atoms which had absorbed them: they produce the *bright-line spectrum*.

The three spectra shown in *Figure 1* are the basic types. However, occasionally one finds a spectrum which has BOTH a continuous part (or a dark-line spectrum) **and** a BRIGHT-line part, as in *Figure 2*:

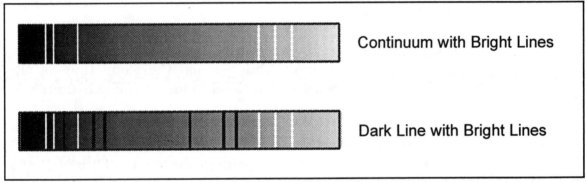

Continuum with Bright Lines

Dark Line with Bright Lines

Figure 2

Since the spectra of *Figure 1* are the basic types, the spectra in Figure 2 must be *unusual* in some way. They are, in fact, COMPOSITE spectra: the light from at least two different things contributes to each of them.

Some fluorescent lights show the continuum with bright lines; the thin gases in the tubes produce the bright lines, while a powder coating the insides of the tubes produces the continuum.

The dark-line spectrum with bright lines is sometimes found for stars which have suffered some sort of *explosion*, throwing hot gases out to great distances; one is seeing light from the star (the dark-line spectrum) as well as bright lines from the distant, glowing gases. Astronomers are always happy to find spectra like these, because they show that something <u>interesting</u> has happened or could be getting <u>ready</u> to happen!

Prisms can disperse white light, but astronomers also use devices called **gratings** for this purpose. There are two types of gratings: *reflection* gratings and *transmission* gratings. Professional astronomers mostly use reflection gratings, which have a metal surface onto which parallel rows of tiny, wedge-shaped furrows called *grooves* have been inscribed. A typical grating has thousands of grooves PER INCH along its surface.

Research-quality reflection gratings are chosen for the straightness and freedom from defects of their grooves and can cost hundreds or thousands of dollars. However, transmission gratings are very inexpensive and are available from many science material suppliers. A transmission grating disperses white light passed through it rather than reflected from it. Even if transmission gratings are not available, the surface of any compact disk (CD) is a low-cost alternative that you can use to produce spectra.

Procedure:

To use a transmission grating, simply look through it at a light source; you should see the light source itself and a spectrum <u>on either side</u>. (If you do not see the spectra, rotate the grating 90° and look again.)

The use of a CD for viewing spectra with the eye) is shown in *Figure 3*. Holding the CD at an angle which will produce a narrow horizontal spectrum across it **does** require some practice.

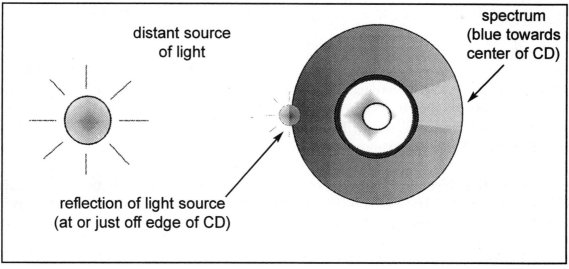

distant source of light

spectrum (blue towards center of CD)

reflection of light source (at or just off edge of CD)

Figure 3

Hold up the CD with the shiny (unlabeled) side facing you. Angle it a bit toward a light source which is moderately distant and not too large (such as a street light or the Full Moon). Put the light source's reflection at, or just beyond, the far left edge of the CD and look toward the opposite side for its spectrum, which will have blue toward the center of the CD and red toward the outer (right hand) edge.

Exercise 15: Spectra with a Grating or Compact Disk

Ideally, you want to have the spectrum appear as a nearly horizontal streak across the CD. The spectrum may appear elsewhere, even in circles around the center; if so, keep trying: you are not yet holding the CD at the proper viewing angle.

If the light source is very large, the spectrum it produces will be too broad and complex (too many overlapping parts) for you to easily interpret it. Stick to bright but distant light sources (such as street lights), or to light sources which are actually small in size (such as candle flames). Try to keep your light sources *similar in apparent size*, by changing your distance from them.

Try to examine only one kind of light source at a time. If you are in a location with many light sources (such as downtown or in a shopping mall), each will contribute its own spectrum, filling your view with colored streaks; you may not be able to identify which one was produced by the particular light source you wanted to study.

Observations:

On one or two evenings, walk around the campus or town with your grating or CD, examining various light sources and their spectra. Try to avoid touching the grating or CD surface as much as possible: hold it by the edges.

Observe and record the spectra of at least six different light sources. As you work, keep a sketch book of your observations. Specify what you observed and where you were.

Draw the spectra as you see them; if there are bright lines, make a note of their particular colors, and if there are dark lines indicate at what color of the continuum they were seen. (Note: Some lights (such as traffic signals) have colored filters in front of them. This will diminish some of the colors you may see. Keep track of any filters in your report.)

Later, in your lab report, reproduce your spectra in color, if possible, using colored pencils or similar media. Explain what would CAUSE each type of spectrum that you observed (*e.g.*, glowing metal filaments, hot thin gases, pressurized gases).

Some possible light sources appear in the following table. Try especially to examine those in bold letters!

The Full or nearly-Full Moon
Street lights: • **Bright orange lights** (high-pressure sodium)   • Bright deep orange lights, usually very long bulbs, seen near highway interchanges (low-pressure sodium)   • **Bright white or bluish-white lights** (mercury vapor)
Yard lights
Stadium lights
Fluorescent lights: Long tubes in ceiling fixtures, or bulb-sized tubes used in place of ordinary (tungsten filament) light bulbs
Tungsten filament light bulbs, including lamp bulbs and flashlight bulbs
Neon signs in windows of department stores and restaurants, and "Power On" lights, as seen on some appliances
Candle flames
Various salts heated over a Bunsen burner. If you have access to a chemistry lab, block off all outside lights and turn off overhead illumination.

 See the LAB REPORT CHECKLIST for guidelines in preparing your report.

Exercise 15: Spectra with a Grating or Compact Disk

Observations of the Setting Sun

You Will Need:
- Protractor
- Several pencils
- Dowel rods or other straight sticks
- Paper
- Two or three observations, spaced 5-7 days apart

Purpose: To observe changes in the horizon positioning of the setting Sun

Background:

Although it rises and sets because of Earth's rotation, the Sun does not follow the exact same path across the sky on each day. As Earth orbits around the Sun, the Sun's position *relative to the stars* moves eastward by about 1° per night, meaning that constellations rise about 4 minutes earlier every night (4 minutes is 1/360 of a day; 1° is 1/360 of a full circle). Also, because Earth's orbit is TILTED about 23.5° to Earth's equator, the orbital motion causes the Sun to drift back and forth between Earth's northern and southern hemispheres: from its low point around December 22, it rises ever higher in our sky until reaching its highest altitude at noon around June 22.

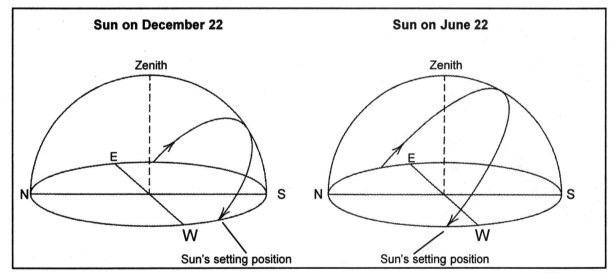

Figure 1

Figure 1 illustrates the extremes in the Sun's rising/setting path across the sky of an observer who is at about 40° North latitude. The height of the Sun at noon, and its rising and setting locations, change significantly. The greatest *change* in the Sun's rising and setting locations occurs around one of the **equinoxes** -- the vernal (spring) equinox (around March 21) or the autumnal equinox (around September 23) -- for it is then that the Sun is moving northward or southward most rapidly.

Exercise 16: Observations of the Setting Sun

Observations:

Find a <u>level location with a clear western horizon</u> and soil into which you can place pencils, dowel rods or other straight sticks. *You must be able to leave the sticks there for a few days, or at least be able to accurately relocate the places you put them into the ground.*

On a clear afternoon shortly before sunset, place one pencil or stick vertically into the ground (call this "stick A"). AS THE SUN'S LOWER EDGE TOUCHES THE HORIZON*, insert ANOTHER stick ("stick B") about two feet in front of stick A so that its shadow is cast onto stick A (see Figure 2); you then know that a line through the two sticks points toward the place on the horizon where the Sun was setting.

> Be sure to note the date and TIME of each sunset. The times should be determined from an accurately-set wristwatch, NOT from published times of sunset in the newspaper or other media source.

On a clear afternoon several days later, return to your observing location shortly before sunset. Find the places where you had put your sticks into the ground (or where they still are, if you were able to leave them in place). Re-insert the sticks if you had to remove them after your first observation. Again, as the Sun sets, place a stick ("stick C") into the ground about two feet in front of stick A, so that its shadow falls onto stick A, and note the date and time accurately.

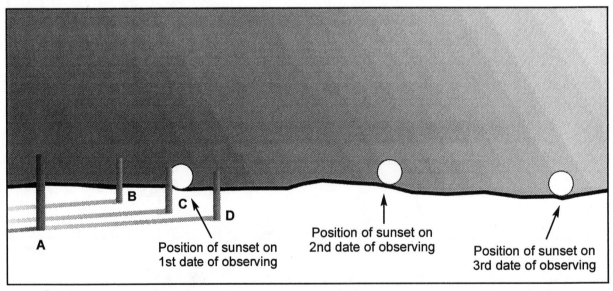

Figure 2

Devise a method (such as thread and protractor) to measure the HORIZONTAL ANGLE (on the ground) between sticks B, A and C (in that order: angle BAC) as accurately as you can. Averaging several estimates is advised! Record the measurements and their average. Also record the *decimal number of days elapsed*

Exercise 16: *Observations of the Setting Sun*

since the first observation, using the fact that there are 24 hours or 1440 minutes in one day.

> **For Example:**
>
> Between Sept. 24 at 5:59 p.m. and Sept. 27 at 5:50 p.m. is 2 days, 23 hours, 51 minutes; this becomes 2 days + (23/24) day + (51/1440) day , or **2.994 days**.

If you can, go out again after another 5 to 7 days and place another stick ("stick D") in line with stick A as the Sun sets. Again measure an angle: this time, angle DAC. Also, again accurately record the date and time, and calculate the elapsed time in decimal days between that afternoon and the one on which you previously observed.

Finally, for each pair of afternoons, divide the number of degrees by the decimal number of days elapsed to get the change (degrees per day) of the Sun's setting position.

Questions:

Is the setting position of the Sun shifting?

Is it shifting toward the right (north) or toward the left (south) as time passes?

Is the number of degrees moved per day constant <u>within the accuracy of your measurements</u>? (You will need the "stick D" measurement to answer this).

Is the TIME of sunset always the same, or do you observe a change for it?

 See the LAB REPORT CHECKLIST for guidelines in preparing your report.

---

* DO NOT look at the setting Sun for more than a second or two -- it is still bright enough to cause eye damage after prolonged viewing. [Quick glances should be all right. Sunglasses (especially those which block ultraviolet light) will provide some additional protection.]

Exercise 16: Observations of the Setting Sun

The Moon's Phases

You Will Need:
- Straight lengths of wood, plastic pipe or metal
- Twelve-inch ruler
- Protractor
- Three short bolts
- Two square nuts and one wing nut
- Brief observations every few afternoons for two weeks, beginning if possible a few nights after New Moon

Purpose: To investigate the relative positions of Sun and Moon in our sky at different phases of the Moon

Background:

As the Moon orbits around the Earth (which is, itself, moving around the Sun), we see the Moon lighted from different directions so that its appearance changes noticeably from night to night. This is the Moon's cycle of PHASES (see *Figure 1*).

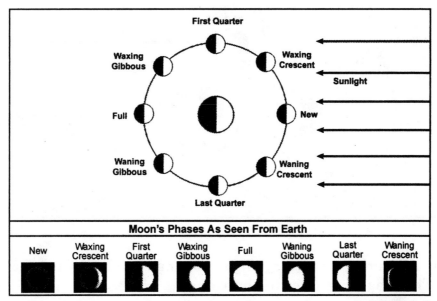

Figure 1

In *Figure 1*, the Sun is far to the right, well out of the view. Its rays of light always illuminate half of both the Moon's and the Earth's spherical bodies. When Sun, Moon and Earth lie nearly along the same line (in that order), sunlight strikes the part of the Moon facing away from Earth, so we look toward the unlighted portion (usually, this means we see nothing, except when the alignment is so exact that the Moon *eclipses* the Sun); the Moon's phase is **New**, meaning a new cycle of phases has started.

For the next two weeks, the illuminated portion of the Moon increases, so phases in that part of the cycle are sometimes called *waxing* (growing) phases; they include **Waxing Crescent**, **First Quarter**, and **Waxing Gibbous**.

(Please note: the Moon's phase is considered to be a Waxing Crescent anytime after New but before First Quarter; it is considered to be Waxing Gibbous anytime after First Quarter but before Full.)

The Moon reaches **Full** phase when Sun, Earth and Moon (in that order) are in a line; sometimes, Earth's shadow can fall onto the Moon then to produce a lunar eclipse, but usually the alignment of the three bodies isn't that perfect. After Full phase, the Moon's illuminated portion shrinks; this part of the cycle contains its *waning* phases.

What determines the phase of the Moon we see is the angle between the sunlight and our viewing direction: the **phase angle**. The phase angle is nearly 180° for New Moon and about 0° for Full Moon; see *Figure 2*.

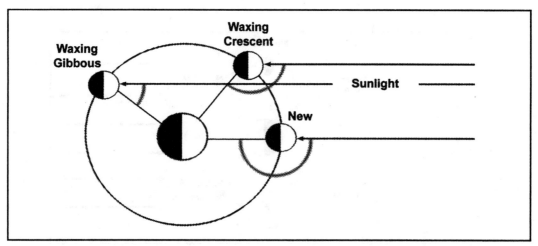

Figure 2

You will be measuring a different angle: the angle in the sky between the Sun and the Moon. That angle is *related* to phase angle:

Phase Angle = 180° *minus* angle between Sun and Moon

Procedure:

You will construct the device shown in *Figure 3* to measure the angle A between the Sun and Moon at times when BOTH are visible in the sky. The device is a variation of an ancient sighting tool called a **triquetrum**, used by Copernicus in the 16th Century. Because the arms (linear portions) of this device are rather long, you

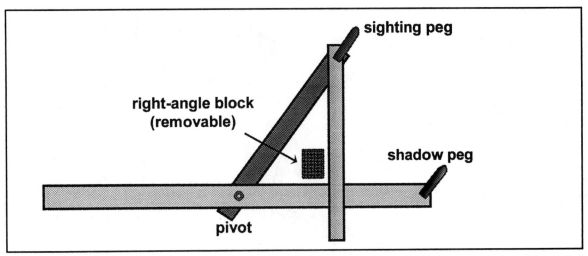

Figure 3

should construct them out of fairly sturdy materials to minimize their bending during use.

The long arm of the triquetrum should be at least two feet long, while each of the other arms should be about 1.5 feet long. The pivot should be a bolt fastened on the underside with a wing nut; this will allow you to tighten the two connected arms in place after making an observation. The sighting peg and the shadow peg should also be bolts, mounted upside down (as shown) and held in place with regular, square nuts.

A small, square block of material (about the size of a child's alphabet block) will be used to ensure that the long arm and the short, hanging arm of the triquetrum are at right angles to each other.

Measure the distance in millimeters between the pivot and the sighting peg; call it "L." Record the value of L for later use.

Once you have constructed the triquetrum (provided it is fairly sturdy), its use is fairly simple (see *Figures 4* and *5*). BOTH SUN AND MOON MUST BE VISIBLE TO USE IT.

With your eye near the pivot, aim the shadow peg toward the Sun but **do not look directly toward the Sun!** Rather, *turn the long arm until the shadow cast by the shadow peg lies along the long arm*. The long arm will then be pointing directly at the Sun.

Now, look from the pivot along the short arm with the sighting peg. Move the short arm about the pivot, and rotate the triquetrum around the long arm, until the

Exercise 17: The Moon's Phases

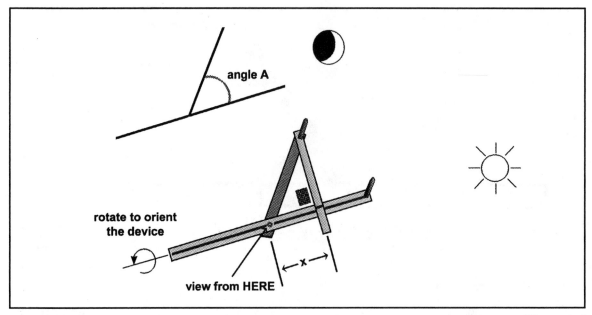

Figure 4

sighting peg is aimed toward the center of the Moon. The shadow cast by the shadow peg should still be aimed along the long arm. Tighten the wing nut to lock both arms into position.

The third arm hangs loosely and will intersect the long arm somewhere. Move it left or right along the long arm until the intersection looks like a right angle. *Use the removable square block* to **make sure** that the two arms cross at a 90 degree angle. Using thumb and forefinger, pinch the long arm and the third arm together at their crossing point so that they do not move.

With a ruler, measure the distance X (in millimeters) between the long arm pivot and the *center* of the third arm where it crosses the long arm; you have already used the removable square block to make sure that the two arms cross at a right angle.

As long as the third arm and the long arm are perpendicular, the angle A between Sun and Moon is easy to find; it follows the relation

$$\text{cosine } A = \cos A = X / L$$

Figure 5 shows how to use the device for Sun - Moon angles greater than 90 degrees. In **this** case, the angle you need (angle B) is just 180 degrees *minus* angle A.

Exercise 17: The Moon's Phases

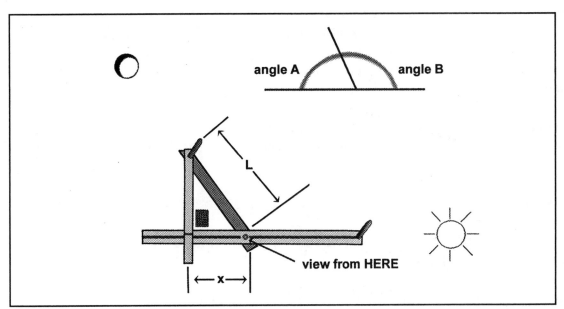

Figure 5

For Example:

Say L = 400 mm and the Moon is waxing gibbous, as in Figure 5.
If you measure X = 260 mm, then cosine A = X / L = 260 / 400 = 0.650

Angle A = arc-cosine (X / L) = arc-cosine (0.650) = 49.5 degrees
Angle B = 180 degrees - angle A = 180 degrees - 49.5 degrees = 130.5 degrees

The phase angle is 180 degrees minus angle B, which here is the same as angle A ! (Phase angle = 49.5 degrees)

Observations:

Consult a calendar, almanac or newspaper for the dates of the next New and Full Moons. Your observations should be made between those dates, starting a night or two after New. (Waxing crescents will be hard to see except around sunset, but First Quarter and waxing gibbous phases are bright enough to be seen earlier in the afternoon.)

On each of three or four afternoons between New and Full Moon, estimate the angle between Sun and Moon with your triquetrum. From those angles, compute the phase angles for those dates.

Exercise 17: The Moon's Phases

Make a sketch of the Moon's appearance on each occasion. Also, note the date, time, weather conditions, and the direction and altitude of the Moon each time.

Consult newspapers or an almanac for the times of local sunset on the dates of your observations.

Objects move across the sky (rising and setting) at a rate of about 15 degrees per hour; if the Moon were 15 degrees <u>east</u> of the Sun in the sky, it would set 1 hour <u>later</u> than the Sun.

Questions:

How is the Moon's phase (as sketched by you) related to the angle between it and the Sun?

How is the Moon's phase related to the Moon's phase angle?

How much later than the Sun would the Moon have set on each of the dates you observed it? Compare your estimates with the Moon's <u>phase</u> (as sketched) and <u>phase angle</u>.

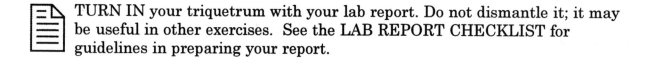 TURN IN your triquetrum with your lab report. Do not dismantle it; it may be useful in other exercises. See the LAB REPORT CHECKLIST for guidelines in preparing your report.

Shadow Stick Astronomy

You Will Need:
- Observations on one sunny day around noon
- One brief observation on a following clear night
- Wristwatch
- Sharpened pencil or nail
- Protractor
- Foot-long ruler
- Graph paper
- About 2 feet of string or other similar material

Purposes: To construct and calibrate a simple sundial, and to estimate the Sun's altitude at noon

Background:

The earliest clocks were sundials, followed by water-drip clocks called *clepsydrae* (literally, "water thieves" [*kleps* (thief) + *hydro* [water]), then weight-driven geared pendulum clocks, then motor-driven clocks, and finally the electronic, battery-powered clocks we use today (not to mention atomic clocks). With these technological advances, our ability to keep track of intervals of time has become ever more precise.

Why do the hands on a clock move clockwise? The answer to this seemingly absurd question is that clock hands mimic the movement of the <u>shadow</u> cast by a vertical stick as the Sun rises and moves across the Northern Hemisphere sky. Originally, time was measured by the altitude of the Sun, estimated from the length of the shadow cast by a vertical stick.

On any given day, the shadow is shortest at the instant the Sun is highest in the observer's sky: *local noon*. In the Northern Hemisphere, local noon occurs when the Sun is due South.

Procedure:

The vertical, shadow-casting part of a sundial is called the *gnomon* (Greek for "indicator"). Usually, the gnomon's shadow falls onto a curved hour/minute scale, but for this exercise, you will make do with a flat piece of cardboard or other stiff material onto which the cardinal directions (N, E, S, W) have been drawn.

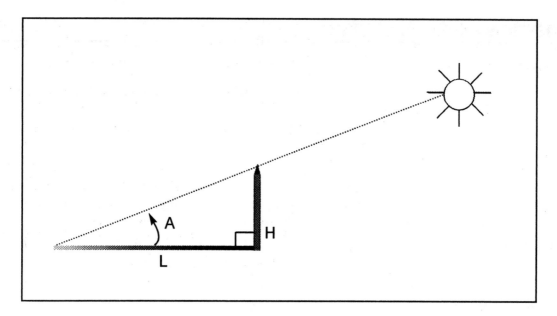

Figure 1

See *Figure 1*. At local noon, when the Sun is due South (for observers in the Northern Hemisphere), the Sun's altitude A (in degrees) and the length L of the shadow cast by a vertical gnomon of height H are all related by:

$$\text{tangent of A} = \tan A = H / L$$

> **For Example:**
>
> If H = 100 mm and A = 45°,
> tan (45°) = 1.000, so L = 100 mm
>
> If H = 100 mm and A = 30°,
> tan (30°) = 0.577, so L = H/0.577 = 1.73H = 173 mm

Since the Sun's altitude is LOWER before noon and also after noon, the gnoman's shadow will be <u>longer</u> at those times.

Refer to *Figure 2*. You will use a sharpened pencil, a nail, or some other straight, <u>pointed</u> object as the gnomon. Choose a piece of cardboard, foam board, or other rigid flat board which has a length <u>three to four times the height of the object you chose as the gnomon</u>. About 1 inch in from an edge and parallel to it, use a ruler and pencil to draw a straight line (this will be your east-west line). Midway along that line, mark a dot as the location of the gnoman.

Now, using a protractor and a long ruler, <u>carefully</u> draw a straight line through the origin dot which is **perpendicular** to the straight line you drew before; this will be your north-south line.

Exercise 18: Shadow Stick Astronomy

Carefully mount the gnoman on (or through) the board at the location of the dot, so that it is vertical: perpendicular to the flat board [**an IMPORTANT step!**]. Measure and record the height (H) of the gnomon, in millimeters, as mounted.

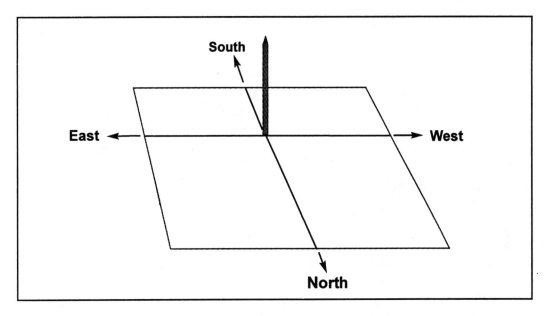

Figure 2

How would you ensure that the gnomon is, and will remain, vertical? Be creative, and explain how you did it in your lab report.

Once you have constructed your sundial, find an outside location which is level, and on which you can make locating marks (such as chalk marks on your driveway, or pegs in the grass).

Observations:

Choose a sunny morning for the first part of the experiment. Using the time signal (on the hour) of a local radio station or some other time indicator, *set your wristwatch to the accurate current time*. Go outside about an hour before local noon (or even earlier, if you can spare the time) and set up your sundial so that the straight line along which the gnomon is mounted (the one drawn 1" from an edge of the board) is facing approximately east-west (see *Figure 2*).

> Do NOT use *Polaris* to orient your sundial! **How else could you establish approximate horizon directions (N, E, S, W) to do this part of the experiment?** What did you do? Include the explanation in your lab report.

Exercise 18: Shadow Stick Astronomy

The gnomon will cast a shadow onto some part of the sundial. **Every five minutes**, make a light, precise mark on the board <u>at the location of the pointed end of the shadow</u> (this location will be changing). Next to each mark, lightly jot down the TIME (from your watch, to at least the nearest minute) of that observation. Continue these observations until around 1:00 p.m. (or even longer, if you have the time), so that you have **at least 20 different shadow locations marked**.

WITHOUT MOVING THE FLAT BOARD, use a ruler to measure the distances (in millimeters) from the gnomon to each point you marked, record these numbers and the TIMES of observation together as a data table. Construct a graph on which you plot shadow length (vertical axis) against time (horizontal axis). FIT A CURVE through all points, and from that curve determine:

(1)   the length of the shadow when it was shortest, and;
(2)   the TIME when the shadow was shortest (see *Figure 3*).

MARK that location (which will most likely fall between two of your observed shadow points) on the sundial.

 CAUTION: DON'T MOVE ANYTHING YET!

Stretch a long piece of string or thread across the sundial so that it passes alongside the gnomon and runs parallel to the direction between the center of the gnomon and the location you marked for the shortest shadow point. Fasten both ends of the thread to the ground with pegs, or mark a sharp chalk mark at each end.

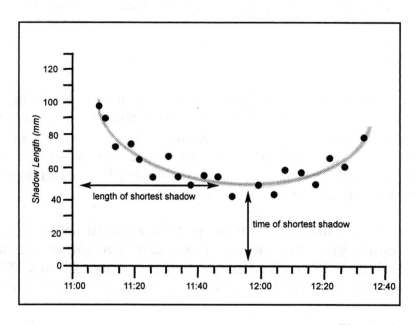

Figure 3

Exercise 18: Shadow Stick Astronomy

The line you have just fastened down or marked is your best estimate of the true north-south line for your location. **How accurate do you think this estimate is?** You will be returning to this location on the next clear night (ideally, on the evening right after your "noon" observations), so you need to make sure that the line or the marks are still there later!

Now you may remove the sundial! HAND IN the sundial with your lab report.

The length of the shortest shadow cast by the sundial (*see Figure 3*) is "L" in the formula tan A = H/L. "H" is, of course, the height of the gnomon, which you have measured. In this case, the angle A is the <u>altitude of the noon Sun</u> on your date of observation.

Calculate the value of A and record it.

How much uncertainty in the time of local noon do you feel you had? Did local noon (as determined from your measurements) **occur right at 12:00 p.m. watch time?** If "sundial noon" was different from "watch noon," write down the difference [watch time *minus* sundial time].

On a clear night after your daytime observations, return to your observing site. Relocate your north-south line. Face toward what you estimate to be North. Locate the Pole Star, *Polaris* (the easiest way is to use the Pointer Stars in the Big Dipper). Is it right above your north-south line? Try to estimate how far to either side of your line is (a cross-staff might help; also, see the exercise "Altitude and Azimuth.").

If Polaris is not right above your north-south line, what might be the reason for its difference in location?

 See the LAB REPORT CHECKLIST for guidelines in preparing your report.

Exercise 18: Shadow Stick Astronomy

Observations of Jupiter's Moons

You Will Need:
- Brief observations over 5-6 nights.
- Binoculars of at least 7x magnification, or a small telescope
- Steady support for binoculars or telescope
- Pencils and sketch paper
- Planisphere (optional)

Purpose: To investigate the motions of Jupiter's four largest moons

Background:

Galileo Galilei was the first scientist to use a telescope for astronomical observations. In 1610, observing the planet Jupiter through his spyglass, he discovered four bright, star-like objects. Night after night, as Jupiter drifted through the patterns of stars, those four bright points of light drifted with it. Sometimes, they would change their relative positions (*see Figure 1*), and sometimes only three would be seen, but these objects were always associated with Jupiter. In Galileo's day, these were controversial observations, because they demonstrated that Earth was not the only center of motion in the cosmos.

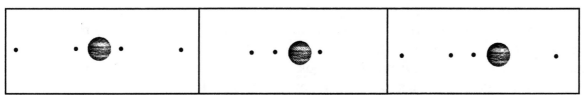

NASA and NSSDC   **Figure 1**

We now know that the four bright objects associated with Jupiter are its four largest moons: Io, Europa, Ganymede, and Callisto. They are often called (collectively) the *Galilean satellites*, in honor of their discoverer. The first (inner) two bodies are about 10% smaller than Earth's own Moon, while the remaining (outer) two are about 10% larger. All four moons orbit essentially in Jupiter's equatorial plane. Io completes an orbit in 1.77 days, while Europa requires 3.55 days, Ganymede takes 7.16 days, and Callisto needs 16.69 days.

Occasionally, one of the moons will disappear (from our perspective) behind Jupiter: an event called an "occultation." Also, occasionally, one of the moons may pass in front of Jupiter (a "transit"), and shortly thereafter observers with moderate-sized telescopes may see that moon's shadow as a black dot on Jupiter's disk. When all four moons are visible, they may be strung out on a line on one side of Jupiter, but

Exercise 19: Observations of Jupiter's Moons

it is not always easy to tell which is Io and which is Callisto (say) because we are observing (essentially in two dimensions) what is a three dimensional situation (*see Figure 2*).

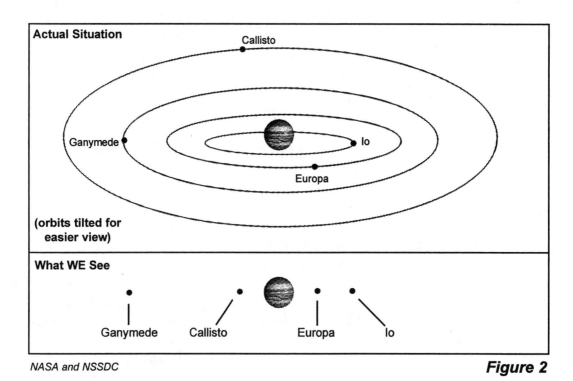

NASA and NSSDC   **Figure 2**

Observations:

Determine, from an almanac or a newspaper or your instructor, in what constellation Jupiter is visible. A planisphere will help you determine when that constellation is visible and where to look. Over several nights, observe the planet from a dark location with binoculars or a small telescope.* To do this exercise successfully, you MUST find some means to hold the binoculars or telescope <u>steady</u>. Telescopes usually have their own mounting system; for binoculars, a tripod may help, or they may be supported on a railing, box, or car top and tilted with the aid of books or a block of wood.

FOCUS your binoculars or telescope on the images of stars in that part of the sky; the stars' images should be as point-like as possible. Now look towards Jupiter. It will take some practice to see the moons, but they are visible even with modest equipment *as long as the equipment is held steady*. They will not be far from Jupiter itself, and you may only see one or two (or NONE!) at first; but, as long as the images are in focus, you will improve with practice-practice-practice!

---

* See the exercise "Apparent Motion of a Bright Planet" for another way to use your observations!

Exercise 19: Observations of Jupiter's Moons

As Galileo did, keep a sketch log of your observations. Sketch Jupiter's disk (which will also give your drawings a sense of scale) and ONLY the moons you can <u>clearly see</u>! (Do NOT assume that all four will be visible on any given night: they are harder to spot when very near Jupiter, or one may be behind the planet.). *If there are any bright stars in the same field of view, sketch their locations relative to Jupiter.* As always, record the date, location, time of night, weather, and any other pertinent information for each observation.

Comment on the position changes of Jupiter's moons.

Question:

Can you tell which moon is which from your observations? Several nights' observations, and the orbital periods of the moons, may allow you to decide. Justify your conclusions.

 See the LAB REPORT CHECKLIST for guidelines in preparing your report.

Observing Sunspots by Projection

You Will Need:
- 4 or 5 clear mornings within a two week timespan
- Binoculars or telescope
- Tripod or other support
- White paper
- Support for the paper
- Large piece of cardboard
- Ruler
- Drawing compass
- Tracing paper

Purpose: To obtain information about the Sun's rotation by mapping the changing positions of sunspots on the projected disk of the Sun

Background:

Our Sun is a rather average yellow star composed mainly of very hot hydrogen and helium gases held together by the Sun's immense gravitation. Like many other stars, the Sun also has a magnetic field, in overall (global) strength not too different from Earth's.

However, most of the time one can find dark spots or groups of spots on the Sun's surface which have magnetic fields much stronger than the Sun's overall field; these *sunspots* are local eruptions of magnetic field from the Sun's interior (*see Figure 1*).

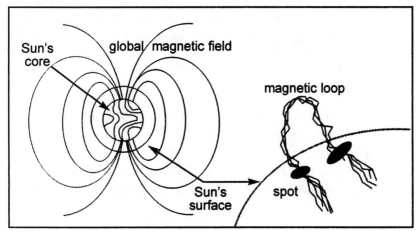

Figure 1

Because it is made of pressurized gases (and not solid materials), the Sun does not rotate at a constant speed; the speed varies with latitude on the Sun and also with depth. This "differential rotation" tends to wind up and concentrate the Sun's interior magnetic field lines. Occasionally, a loop of concentrated magnetic field lines pushes its way upward from the interior, out through the Sun's surface and beyond. At the locations where this loop intersects the Sun's surface, sunspots are seen. Though not "cold" by any means (they are roughly the temperature of red-hot iron), sunspots do not glow as strongly as the rest of the Sun's surface because the dense magnetic field that forms them inhibits the normal rising motions of gases which carry heat upward from the Sun's interior.

In the early 17th Century, the scientist Galileo first used a telescope to create a projected, magnified image of the Sun on which he could observe sunspots. Galileo noticed that the positions of the spots on the Sun's disk changed from day to day, and he correctly concluded from this that the sun was a rotating body (*Figure 2*).

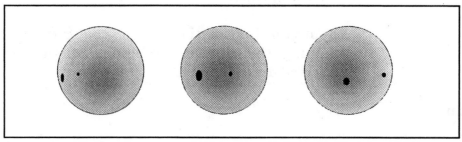

Figure 2

Procedure: In this experiment, you will observe the image of the Sun as projected onto white paper or other white material by binoculars or a telescope, to look for sunspots and map their positions on different mornings.

 NEVER look at the Sun directly through any optical equipment; you will suffer permanent eye damage. Projection is a safe way of observing the Sun because you do not view it directly.

Refer to *Figure 3*. If you are using binoculars, you will need some sort of support for them, such as a tripod with a swivel (pan) head. If you do not have a binocular tripod mounting post (a claw-end affair that clamps over the central shaft of the binoculars and screws onto the tripod), try fastening the binoculars to the tripod head with wire, rubber bands, or even masking tape. Most small telescopes already have tripod mounts.

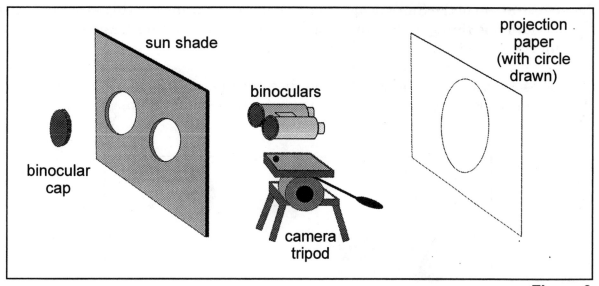

Figure 3

Remove the lens covers from the binoculars, and set them down in the middle of a large (at least 8 by 10 inches) flat piece of cardboard. With a pen or pencil, trace around the two binocular ends, then remove the binoculars and cut out the circles just traced on the cardboard. The sheet of cardboard has become your *sun shade* and will be mounted over the binoculars when you are observing (a telescope will only have one hole).

Aim your binoculars or telescope toward the Sun, but **do not look through it**. The best way of aiming is to look at the shadow cast by your equipment: when the shadow of your binoculars or telescope tube is shortest and roundest, you are aimed directly at the Sun and should see light coming through the eyepieces.

 Do not look through the eyepiece at any time!

Once you see light coming through the eyepiece, mount the sun shade. Cover one of the binocular lenses with its cap (only one lens is needed for projection.). Project the Sun's image onto a sheet of white paper that you support from the back. Move the paper away from the optical equipment until you get the **largest** image of the Sun that you can *focus sharply*; this should be 1-2 inches in diameter at a projection distance of around three feet for binoculars, larger (because of their greater focal lengths) for telescopes.

Use a drawing compass to make a circle of similar diameter in the middle of the sheet of white paper. Repeat this for other sheets, or make several copies of your first sheet; the sheets will be used to record your observations of the Sun over several different mornings*.

Exercise 20: Observing Sunspots by Projection

The **orientation** of the Sun's image is particularly important for this exercise**. The North-South line through the Sun's center will be seen at different amounts of tilt as the Sun rises (see Figure 4), so it is important to:

1)   Observe the Sun around the same time each morning, and
2)   Hold the white paper in the same way for each observation.

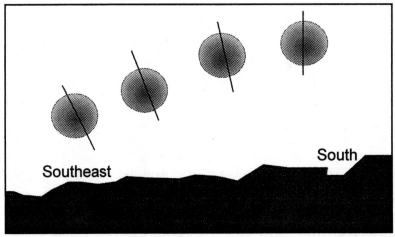

Figure 4

Observations:

At about the same time on each of four or five clear mornings over about a two week timespan, set up your equipment, but **do not** mount the sun shade.

*   *Mornings are the best time to observe the Sun because the Earth's atmosphere is still relatively cool and still. By afternoon, the air has become heated and turbulent, which blurs the Sun's image and makes the projected images of sunspots harder to see clearly.*

**   *See Figure 3 of the exercise "Exploring the Winter Sky."*

Aim your binoculars or telescope toward the Sun, but **do not look through it**. Follow the procedure in the previous section.

Once you are correctly aimed, mount the sun shade. If you are using binoculars, also cover one of the lenses with the lens cap at this time, as you only need one lens.

Aim the white paper (on which you have drawn the circle) so that the Sun's image is projected onto it. Support the paper from behind with a clipboard or book.

Adjust the distance of the paper so that the Sun's image FILLS the circle.

Adjust the tilt of the paper so that the Sun's image is ROUND.

Focus the image of the Sun using the binoculars or telescope focusing knob, so that the edge of the projected image is sharpest. (You may now be able to see some sunspots).

TURN the paper to the orientation (say, vertical) that you will be using each morning.

Use a pencil to carefully mark the locations of any sunspots or groups of spots that you observe on the projected disk of the Sun. **If you see no sunspots, still make a note of that fact!** Carefully record date, time, weather conditions, and other information you normally put into lab reports. **Look away from the Sun frequently to rest your eyes; even in projection, it's very bright!**

Make observations of the Sun every few mornings for about two weeks, then arrange your Sun observing sheets in chronological order (earliest date first) and try to identify particular spots which have changed location between observations.

> If you have taken some care to maintain the orientation of the Sun's image over the course of your observations, you can combine them using tracing paper.
>
> Using a drawing compass, make a circle on the tracing paper which is the same size as the ones on your Sun observing sheets. Place the tracing paper circle over the circle on your first observing sheet, and mark the first day's sunspot locations on the tracing paper. Repeat this using the sheets for the remaining dates.

What sort of path does each spot make in crossing the Sun's disk?

 See the LAB REPORT CHECKLIST for guidelines in preparing your report.

Exercise 20: Observing Sunspots by Projection

Observations of the Moon's Features

You Will Need:
- Binoculars or a small telescope
- Two clear nights for observing: a night when the Moon's phase is near First or Last Quarter, and a night when the Moon's phase is close to Full

Purposes: To become familiar with the major surface features of the Moon and how their appearance changes with the Moon's phase

Background:

Earth is unusual in having a moon which is so large relative to the planet it orbits; the Moon is about one-fourth Earth's diameter. The APOLLO missions returned samples of lunar rocks which show that the Moon's surface is very old: in some places, well over four billion years.

Though the exact origin of our Moon is still not known, one popular theory has it being "born from Earth" as the result of an enormous collision between our planet and another planetary body roughly half Earth's size. While this may sound like science fiction, the chemical similarities of rocks on both Earth and Moon argue for some common origin. Also, the early solar system's development involved the formation of moons and planets from smaller chunks of material (some of which we still observe as asteroids and comets), so it is likely that debris of many sizes was moving around among the young planets and would sometimes strike them.

UCO/Lick Observatory image   **Figure 1**

See *Figure 1*. Even to the unaided eye, the Moon shows a non-uniform appearance: white with a complex mix of lighter and darker greys. Different people imagine that they see a grieving face, or a woman's head in profile, or a leaping rabbit in the Moon's darker regions. (Can you see any of these?)

The **same** lunar surface regions face Earth <u>all the time</u> -- whether the Moon's phase is Crescent, Quarter, Gibbous or Full -- because Earth's *tidal pull* has forced the hemisphere of the Moon now facing us to <u>continue</u> facing us. This means that you can observe the <u>same</u> lunar surface features at <u>different</u> lunar phases and see how they change in appearance.

Whatever the details of its origin, the Moon shows us an old, battered surface which

Exercise 21: Observations of the Moon's Features

has hardly changed in appearance over the last few billion years. Earth is a large planet which has a great deal of internal heat available to reshape its surface by moving about continents and producing volcanic eruptions, and there is also erosion from wind and running water. In contrast, the Moon's gravity is too weak to hold much of an atmosphere (virtually none, in fact), which also prevents water from existing there in a liquid form; and, the Moon' s small size resulted in a quick loss of its internal heat. Compared to Earth, the Moon is a dead world and has been dead for a very long time.

This is not to say that the Moon is a dull object to study! The Moon's surface shows the scars of many ancient events that were also probably occurring on the young Earth: meteor bombardment and eruptions of lava while the body still had its internal heat. Studying the Moon's surface gives astronomers important clues to conditions in the solar system's first billion or so years of existence.

The Moon's surface shows three major types of features: **highlands**, **maria**, and **rayed craters**.

The *highlands* are the oldest regions on the Moon: light grey rock pitted with thousands of meteor impact craters. The most obvious highland regions lie near the northern (top), central, and southern (bottom) portions of the Moon's disk shown in *Figure 1*. The highlands show the effects of the frightful numbers of meteors striking the Moon's surface in the earliest days of the solar system.

The *maria* (a Latin word meaning "oceans;" the singular form is *mare*) are the large, dark, roundish regions. They tend to be younger than the highlands. The maria are actually huge, ancient craters that filled with lava seeping from the Moon's interior during its first billion or so years of existence. The maria tend to be smoother than the highlands because after the Moon's first billion years, most of the meteors (interplanetary debris) had already struck the surface.

The *rayed craters* formed after the Moon's surface had cooled and solidified, so they are relative latecomers. Pulverized rock from these impacts tends to be bright (like chalk) and was flung in all directions, creating obvious, showy "rays" and craters which are themselves also bright.

Why should lunar features change their appearance at different lunar phases? The answer lies in considering the topography of the Moon's surface and the concept of *phase angle**.

Figure 2 illustrates the geometry by which lunar features (in this case, a mountain peak) are illuminated at different phases. The Moon's surface is not flat: it contains mountainous regions and deep hollows with raised rims (the craters). When the Moon is near First Quarter or Last Quarter (phase angles of 90° and 270°), surface features near the center (north-south) line on the Moon's disk are lighted from the

Exercise 21: Observations of the Moon's Features

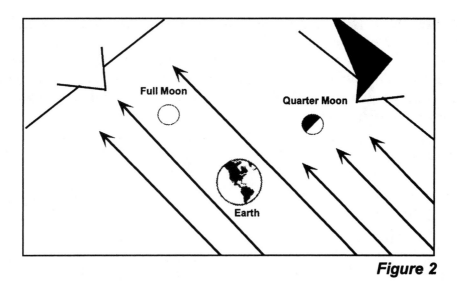

Figure 2

side, which causes them to cast long shadows. The best way to appreciate the *ruggedness* of the Moon's surface is to view it near a quarter phase.

On the other hand, near Full phase (phase angle around 0°) most lunar features are illuminated "straight on" (Sun at a "high noon" position in the sky). Surface features do not cast long shadows under such lighting. However, the pulverized rock in and surrounding <u>rayed craters</u> brightens *dramatically* near Full phase because it is especially good at *reflecting* direct lighting.

Procedure:

Figure 3 is a map of the Moon's more prominent surface features. The Moon is oriented as it is normally seen when due South in the sky. (For a rising Moon, turn the map a bit counterclockwise; for a setting Moon, turn the map a bit clockwise.) Refer to the map while reading the text below.

Lunar craters are named after historical figures (usually scientists and philosophers), such as **Plato**, **Copernicus** and **Archimedes**. The Moon's dark areas (once thought to be actual bodies of water) are generally named after weather conditions associated with water, such as **Mare Imbrium** (Sea of Rains), **Mare Vaporum** (Sea of Mists), **Mare Nubium** (Sea of Clouds), **Oceanus Procellarum** (Ocean of Storms) and **Sinus Iridium** (Bay of Rainbows) or conditions associated with the bodies of water themselves, such as **Mare Tranquillatitis** (Sea of Tranquility), **Mare Serenitatis** (Sea of Serenity) or **Palus Putredinis** (Marsh of Corruption).

The DARKEST areas on the Moon [not counting shadows cast by objects!] include (looking right to left) **Mare Crisium**, the crater **Plato** (at the top of Mare Imbrium),

* *See the exercise "The Moon's Phases."*

Exercise 21: *Observations of the Moon's Features*

Mare Nubium, **Mare Humorum** and **Grimaldi** (far left edge).

The BRIGHTEST areas include (again, right to left) the rayed craters near **Stevinus** (lower right) and the rayed craters **Proclus** (small but bright crater near the left edge of Mare Crisium), **Tycho** (near the bottom), and **Copernicus**, **Kepler** and **Aristarchus** (left of center).

The most MOUNTAINOUS regions are above center, between Mare Serenitatis and Mare Imbrium: the **Apennine** and **Caucasus Mountains**.

The most HEAVILY CRATERED regions are below center, to the right of and below Mare Nubium. Some obvious craters here (especially near Quarter phase) are **Ptolemaeus**, **Alphonsus** and **Arzachel**, lying just below center. The very large craters **Maginus** and **Clavius** are nearly at the bottom of the map.

For detailed observations of this sort, you MUST make sure sure that your optical equipment (binoculars or telescope) is properly supported (preferably on a tripod) and steady.

Observations:

Plan to observe the Moon within 1 to 3 days of either First Quarter or Last Quarter phase**, and again within 1 to 3 days of its Full phase.

Choose ONE bright rayed crater, ONE dark area, and TWO prominent (regular) craters to observe. They must be visible during both nights you observe; for example, the rayed crater Copernicus is visible near Full Moon and also at Last Quarter, but not at First Quarter (it is on the left half of the map). On each night, observe these features with your optical equipment. Sketch each feature and comment upon its overall visibility, especially relative to the features around it.

After making both nights' observations, discuss CHANGES in the appearance of each chosen feature between the two phases you observed.

Questions:

If the feature was a *crater*: Could you see it casting shadows? On which date of observation were the shadows **longer**?

If the feature was a *rayed crater*: When were the rays most prominent? When was the crater itself most easily seen?

If the feature was a *dark area*: When was it most easily seen, relative to the terrain around it?

**   The Last Quarter Moon shows many interesting features, but you should be aware that it is only visible in the "wee hours" after midnight. (This phase is definitely for night owls!)

Exercise 21: Observations of the Moon's Features

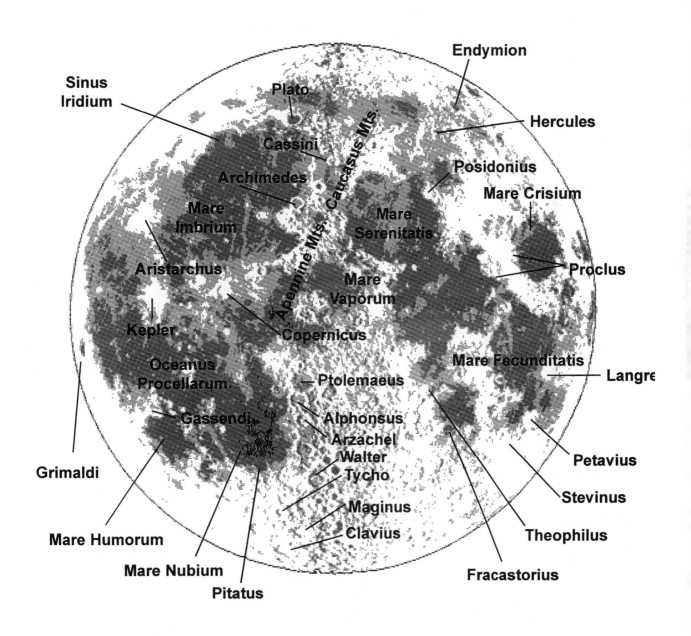

UCO / Lick Observatory image

Figure 3

Exercise 21: Observations of the Moon's Features

Apparent Motion of a Bright Planet

You Will Need:
- Planisphere
- Cross-staff (optional)
- Brief observations every few evenings for two or three weeks

Purpose: To chart the gradual motion of a bright planet through the stars of a constellation

Background:

The sky-watchers of antiquity believed that Earth was fixed (unmoving) at the center of the universe, and that celestial objects such as the Sun, Moon and stars circled about it.

Planets were different. The word "planet" comes from the Greek word for "wanderer." On any given night, a planet will rise and set just like other celestial objects. However, over the course of a few nights, it becomes apparent that a planet also "drifts" through the patterns of stars in a somewhat complicated fashion, as though it has a "will of its own." That is, in fact, a belief that many ancient peoples held: that planets were embodiments of gods and moved through the heavens in their own elaborate fashion.

Today, we know that the complex motions of planets have simple causes. Earth is NOT fixed but orbits the Sun, and Earth's orbital motion has to be taken into account when interpreting a planet's apparent movement through the constellations. Also, the orbits of Earth and other planets are NOT exactly circles but more general geometrical shapes called **ellipses**; the bodies' distances from the Sun (and from each other) vary, which affects the observed amount of a planet's motion against the background stars.

Procedure:

Consult a local newspaper, an almanac like the *Old Farmer's Almanac*, a periodical like *Astronomy* or *Sky & Telescope*, a Web site or your instructor to determine which bright planets are visible and in what constellations they are currently located. Use your planisphere to determine which of these constellations will be visible at a convenient time of night for you to view them, then choose **one** planet to observe.

Because of Earth's orbital motion, which shifts our viewing direction in space about 1 degree eastward every 24 hours, the constellations appear to reach the same location in the sky about *4 minutes earlier every night*. Therefore, to *control the location* of your chosen constellation, it is useful to observe it (and the planet moving through it) 4 minutes earlier for every night that passes after the first one.

For Example:

Sagittarius is due South at 9 p.m. Daylight Time on September 1. It will also be due South on September 2 at 8:56 p.m., on September 3 at 8:52 p.m., and on September 16 at 8 p.m.)

Using the planisphere as a guide, make a good-sized sketch (indoors, before observing) of the brightest stars in your chosen constellation, using larger dots for the brighter stars. It may also be helpful to include some stars from surrounding regions of sky. (Although you will be able to fine-tune your drawing outside, try to accurately portray the star pattern before observing.)

Observations:

Go outside on a clear night and locate your constellation with the aid of the planisphere. Do the relative locations of the stars on your sketch match what you see? If not, make some corrections to your sketch. After the corrections, observe your constellation and add some of its fainter stars to your sketch using smaller dots.

Any bright star that does not appear to be a regular part of your constellation is likely to be the planet! Mark its position in the constellation on your drawing, then *estimate its angular distance* from several bright stars in the constellation, either by "four finger" measurements* or by making measurements with a cross-staff**. Jot down those numbers (identify, in some fashion, which stars you used) and the date and time of the observation.

You will want to make several photocopies of your sketch, for use on subsequent nights. Over a two to three week period, go out every few nights (<u>remember</u>: *four minutes earlier* for <u>each</u> night after the first one!) and repeat your observations and angle estimates.

Reserve ONE photocopy of the constellation field, on which you will plot ALL your planet positions from the different nights (see *Figure 1*).

Exercise 22: Apparent Motion of a Bright Planet

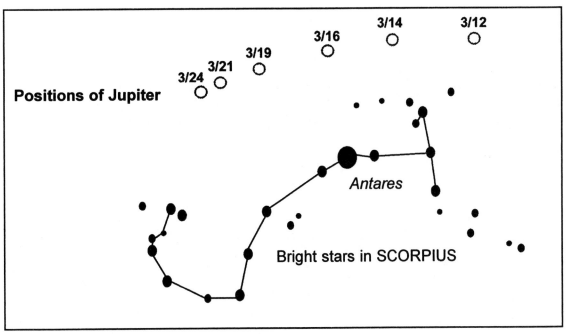

Figure 1

Questions:

Is the amount of the planet's motion the same from week to week?

Is it always in the same direction relative to the constellation pattern?

 See the LAB REPORT CHECKLIST for guidelines in preparing your report.

* See the exercise "Altitude and Azimuth."

** See the exercise "Construction and Calibration of a Cross-Staff."

Exercise 22: Apparent Motion of a Bright Planet

Spectral Classification

You Will Need:
- Calculator
- Metric ruler
- Spectra (provided)

Purpose: To investigate how surface temperatures of stars are determined from the patterns of dark lines in their spectra

Background:

It was Isaac Newton who discovered in 1666 that a glass prism dispersed white light into a spectrum of colors. The study of spectra of stars dates back to 1802, when William Wollaston discovered dark lines in the spectrum of the Sun. Twelve years later, Joseph von Fraunhofer identified hundreds of these dark lines and gave the stronger ones letter names which are still in use today (for example, the H and K lines of calcium, which are especially strong in the spectra of cool stars). In 1859, Gustav Kirchhoff formulated his three laws of spectra: Hot, dense materials (solids, liquids, and pressurized gases) radiate all colors (a **continuous spectrum**), while thin gases show a **bright line spectrum** (only certain wavelengths emitted) when viewed alone. However, a thin gas seen IN FRONT of a hot dense material produces a **dark line spectrum** (also called an absorption spectrum) because the atoms in the thin gas will use some of the continuous spectrum energies to change the energy states of their electrons.

Around 1870, viewing stars through a telescope and prism, Angelo Secchi realized that the spectra of stars could show several different patterns of dark lines and occasionally even dark bands. The simplest spectra, with the fewest lines, were of white stars like *Sirius*. The Sun's spectrum was more complex, with many faint dark lines. Red stars had even more dark lines and/or dark bands (which we now know are really close groupings of many dark lines). In 1885, Johann Balmer showed that the most obvious dark lines in the spectrum of stars like *Sirius* were due to the element hydrogen and had a regular, predictable pattern: the famous *Balmer series*.

The first systematic survey of the spectra of stars was carried out at the observatories of Harvard College, starting in 1886 and continuing into the first decades of the Twentieth Century. Photographs of star fields were taken through huge prisms mounted over the telescopes, so that instead of the usual images of stars, the photographs recorded images of their spectra. The survey eventually sampled the entire sky: well over a quarter-million stars, from the brightest ones down to those which would be barely visible in modern binoculars. It was from this monumental work that the first detailed classification scheme for stellar spectra developed, largely due to the work of Annie Cannon, who categorized most of the line

Exercise 23: Spectral Classification

patterns of a quarter-million spectra using only her eyes and a microscope!

In the *Harvard spectral sequence*, the appearance of the patterns of dark lines was sorted from *simple to complex* and assigned a letter of the alphabet: A, B, C, and D for stars with strong hydrogen lines (Balmer series) but very few others, E through L for stars with the strong pair of H and K calcium lines and lots of others, M and N for stars with lots of lines and dark bands, O through Q for stars with *bright* lines in the spectrum (unusual).

The Harvard sequence ultimately was abandoned in its original form. The reason was that in 1920, physicist Meghnad Saha showed that the patterns of lines were different from star to star mainly because (1) different chemical elements absorb light well at different temperatures, and (2) stars *have* different temperatures. For example, hydrogen absorbs best at around 10,000° Kelvin (18,000°F), while iron absorbs best at lower temperatures and helium does better at higher temperatures. Once this was realized, the classification sequence could be rearranged to show changes from hot stars to cool stars; the modern *spectral type sequence* had emerged. The modern sequence, with letters rearranged and some unnecessary categories eliminated, is:

O------------B------------A------------F------------G------------K------------M
(hottest stars)   (coolest stars)

for example,   *Mintaka*   *Sirius*   Sun   *Betelgeuse*
                (Orion's belt)

This sequence can easily be remembered by the mnemonic:
Oh, **B**e **A** **F**ine **G**irl (or **G**uy), **K**iss **M**e.

To make the sequence a little more accurate, the seven *spectral types* above are each divided into sub-types by adding a number from 0 to 9 to the letter; for example, A2 or M5. The number increases toward the right; e.g., A0, A1, A2...,A9; then F0, F1, F2.... The higher the number, the cooler the star is *within its type*. An A2 star is hotter than an A7 star, but all A stars are cooler than all B stars. Our Sun has the classification G2, making it a hot example of a G-type star.

In the first half of the Twentieth Century, stellar spectra were recorded on glass photographic plates and were, literally, microscopic! Today, the high-technology revolution has reached astronomy and very little direct photography of this kind is still done. Instead, the light from a star's spectrum is directed towards a *charge-coupled device* (CCD): a tiny silicon chip manufactured as a grid of light-sensitive squares, each of which produces a tiny electrical charge when struck by light. The spectrum emerges as a digital signal (an array of numbers) that can be stored on magnetic tape and manipulated later on a computer.

Exercise 23: Spectral Classification

See *Figure 1*, which shows some sample digital spectra of a white (hot), a yellow (moderately hot), and a red (cool) star. Blue light (shorter wavelengths) is towards the left, red light (longer wavelengths) toward the right on each spectrum. The first two spectra especially look like humps with gouges. The overall hump shape is the star's continuous spectrum, while the gouges are the dark lines: places where the atoms in the star's thin atmosphere have removed light.

The most noticeable lines in the white star's spectrum are due to hydrogen: the Balmer series. Notice that these lines have a regular pattern, being more closely spaced toward the left side of the spectrum. The one line that seems to break the pattern is not due to hydrogen: it's the K line of calcium.

In the yellow star's spectrum, the Balmer series is less noticeable, and the two strong H and K lines of calcium are the most obvious features (at the far left).

The red star's spectrum looks chopped up on the right, and this is due to dark bands (bunches of lines); they are due to molecules forming in the star's atmosphere, which immediately tells us that this is a very cool star in comparison with the others. Also, there is a very strong dark line near the middle of the spectrum. This is due to calcium, but it is <u>not</u> the H or K line.

Most of the time, digital spectra are further processed to make them simpler to analyze. What we are mostly interested in is the pattern of dark lines, not the continuum hump, and this often removed by comparing each star's spectrum with the (line-less) continuous spectrum of a tungsten-filament light bulb, or something similar. The process, called flattening, removes the hump, so that we can examine the lines on a level background.

Exercise 23: Spectral Classification

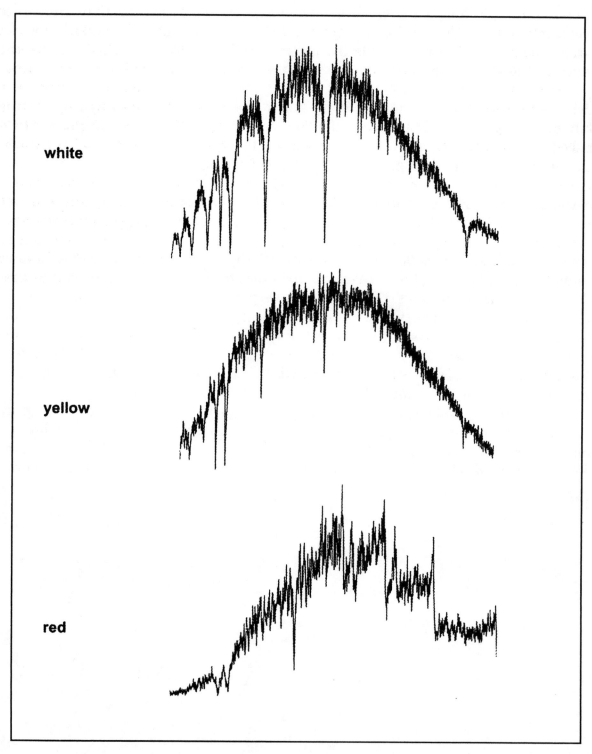

Figure 1

Exercise 23: Spectral Classification

Procedure:

Examine the flattened spectra of stars of known spectral type that appear in *Figures 2* and *3*. The hydrogen lines (labeled "H") and the H and K lines of calcium (labeled "CaII") are marked; a few other lines are also identified. As you compare the spectra, look for CHANGES in the relative strength of two or more lines as you look down the spectra (from hot towards cooler types). One line may weaken (get shallower-looking), another strengthen (deepen), and a third line may hardly change in depth. It is the using of line strength RATIOS in this fashion which allows astronomers to accurately find the spectral type (and sub-type, 0-9) of stars.

Develop your own criteria for spectral typing. You are following in the footsteps of Secchi and Cannon, but with better data! Look for lines changing strength from one spectrum to another. Use a few different pairs of lines when developing strength ratios. Also, look for the appearance or disappearance of some features (such as bands) in the spectrum. The overall appearance of the spectrum (say, a few lines or many lines) might also prove useful. Remember, this is what Harvard College first used.

Once you have developed a scheme for distinguishing one known spectrum from another, see the "unknown" spectra in *Figure 4*. Classify and record them, and carefully explain in your report how you did so.

 See the LAB REPORT CHECKLIST for guidelines in preparing your report.

Exercise 23: Spectral Classification

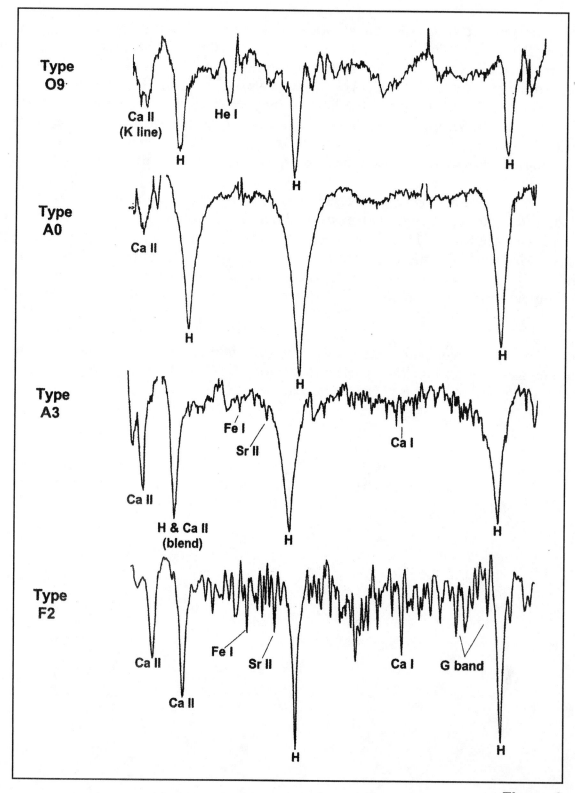

Figure 2

Exercise 23: Spectral Classification

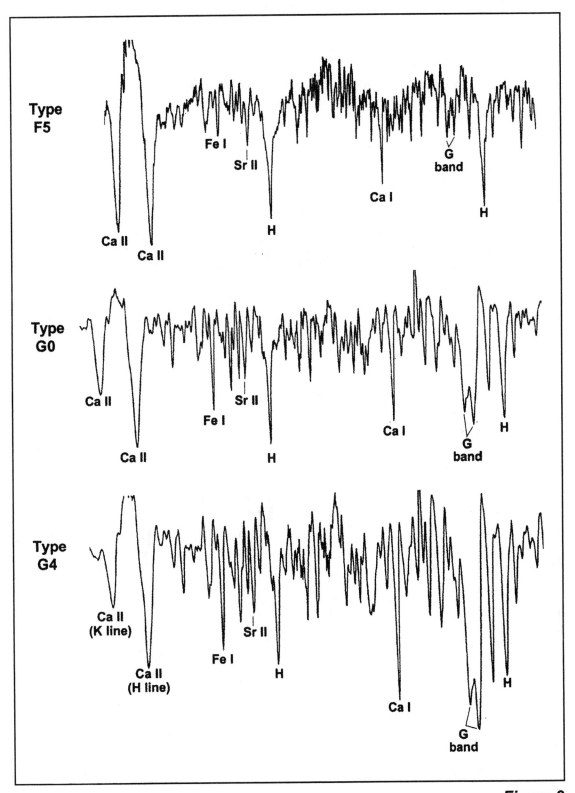

Figure 3

Exercise 23: Spectral Classification

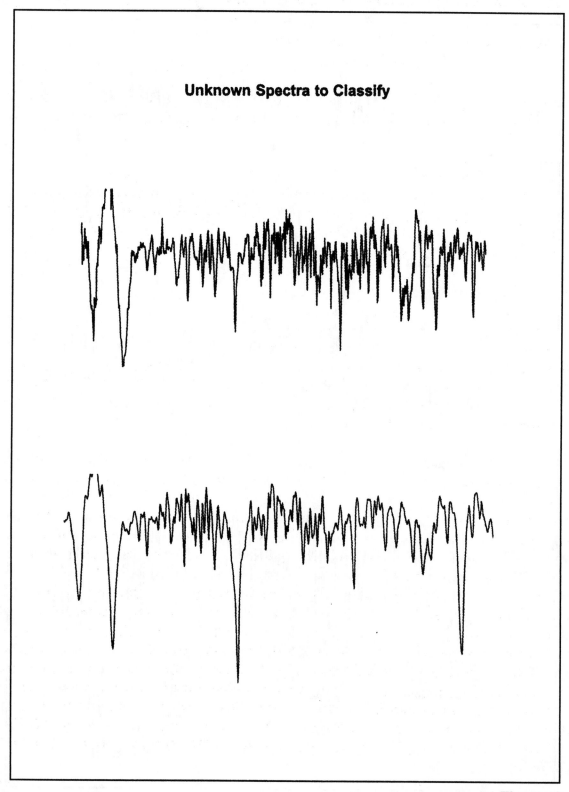

Unknown Spectra to Classify

Figure 4

Exercise 23: Spectral Classification

The Hertzsprung-Russell Diagram

You Will Need:
- A straight edge ruler
- A pocket calculator

Purpose: To investigate how the physical properties of a star are estimated from its location on the Hertzsprung-Russell Diagram

Background:

A star is a volume of heated gases which (except for a thin outer atmosphere) are under very great compression due to the "self gravity" of the star: the tendency for all its particles to attract each other gravitationally. By itself, this tendency would collapse any star. But, a star's gases (heated by the flow of energy through them) also exert pressure (a tendency to expand), and that balances the tendency to collapse during most of a star's existence.

This simple but powerful concept (called *hydrostatic equilibrium* (HSE) by astronomers) is very important for understanding the structure and evolution of stars. It tells us that stars must be hotter in the center than near the surface, for more energy is needed in the center to support the weight of the star. Stars of higher mass (which usually also have a larger size) have greater central temperatures than stars of low mass.

Astronomers did a lot of theoretical and observational work during the 19th and early 20th Centuries to determine the sources of energy that might power the stars. Simple **shrinkage** of all or part of a star would generate heat and therefore gas pressure, but stars would only be able to survive a few tens of millions of years in this way. (The oldest objects in our solar system are meteorites estimated to be almost 4600 million years old.) **Chemical reactions** were suggested as a power source, but chemistry requires atoms, molecules or ions, and very few of these exist deep inside stars; the high temperatures break them into subatomic particles like protons and electrons. As understanding grew of the enormous amounts of energy the nucleus of an atom could store, **radioactive decay** (breakdown of nuclei) and **nuclear reactions** were also suggested. But the *chemical* study of the atmospheres of stars through their spectra showed that stars are mostly made of the two lightest elements, hydrogen and helium. This finding ruled out processes involving heavier elements (such as uranium) and eliminated radioactive decay and nuclear fission as important power sources. We are left with *nuclear fusion* as the primary energy source for stars, as was first proposed in 1926 by the British physicist Sir Arthur Eddington.

For as long as they can, stars maintain a steady glow of internal power. Eventually, however, their power sources begin to wane. Fusion is a difficult process, one requiring extremely high temperatures, during which energy is released while *nuclei* of one element are fused together to make *nuclei* of another. In the Sun, nuclei of hydrogen atoms are fused into nuclei of helium atoms, and energy in the form of gamma rays is released. As an element, helium is even harder to fuse (into carbon and oxygen) than hydrogen, so once the "hydrogen runs out" it is difficult for all but the more massive stars (with hotter centers) to use the fusion of heavier elements as energy sources. STARS EVOLVE because their energy sources are limited, but their self-gravitation (tendency to collapse) is always present.

During the same time as theoretical advances were being made, astronomers also made great advances in determining the fundamental properties of stars. Photography, which came into wide use in astronomy during this time, allowed information on the brightnesses, colors and positions of thousands of stars to be recorded on photographic plates. The light of a star could also be dispersed into a rainbow of colors by passing it through a prism, and a photographic image of its **spectrum** recorded.

As Earth orbits the Sun, nearby stars show minute position shifts (*parallaxes*) when compared to more distant background stars. Photography thus allowed astronomers to determine the **distances** to thousands of nearby stars. A star's energy spreads in all directions, and the amount we receive (*brightness*) is diminished by distance. However, if a star's distance is known, we can correct its observed brightness for the spreading effect and estimate the star's luminosity or actual power level. [See the exercise "Apparent Magnitudes of Stars." **Apparent magnitudes** (symbol: m) are a ranking of stars' brightnesses. A ranking of stars' luminosities produces numbers called **absolute magnitudes** (symbol: M).]

The **surface temperatures** of stars can be estimated in two ways: from their *color index*, and from their *spectral type*.

Color index is a number which compares the brightness of a star in two different color regions, such as blue and red. A hot star will emit more blue light than red light; the opposite would hold for a cool star.

See Figure 1. Beneath a star's atmosphere, its gases are highly compressed. Any heated, compressed substance emits light in a very predictable fashion called a *continuous spectrum**; all the colors from violet to red (and beyond) are represented. But the thinner gases of the star's atmosphere selectively absorb particular, precise colors from this spectrum, so the spectrum we observe for stars is not exactly a continuous spectrum but rather a "continuous spectrum with gaps" caused by the chemical elements in the star's atmosphere. This is called a *dark-line* or *absorption spectrum**.

*   See the exercise *"Spectra with a Grating or Compact Disk."*

Exercise 24: The Hertzsprung-Russell Diagram

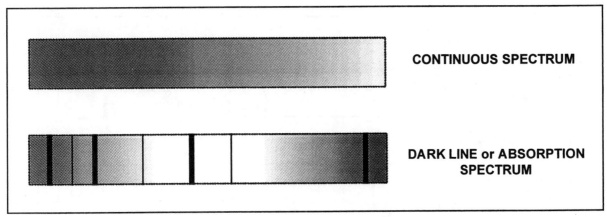

Figure 1

Stars have different patterns of dark lines in their spectra. This was first observed by Angelo Secchi in 1867 and later developed more fully at Harvard through the work of people like Annie Cannon. In the Harvard classification scheme, the spectra were categorized by how complicated the patterns of lines were; the simplest pattern fell into Type A, while a spectrum of Type K (higher in the alphabet) would have a very complicated pattern. In 1920, physicist Meghnad Saha showed that the dark lines were due to different chemical elements in the star's atmospheres, and that the relative depths of lines of different elements were an indicator of surface temperature. Once that became clear, the original Harvard sequence was rearranged and modified so as to reflect the temperature sequence.

The spectral type sequence, running from hottest stars at the left toward the coolest stars on the right, is:

O B A F G K M

Yes, this is one of the stranger things to remember! Fortunately, the American astronomer Henry Norris Russell devised a mnemonic (jingle) to help you:

Oh Be A Fine Girl/Guy, Kiss Me

Stars of Type O are the bluest and hottest stars, while stars of Type M are coolest. *Figure 2* displays some typical spectra.

Exercise 24: The Hertzsprung-Russell Diagram

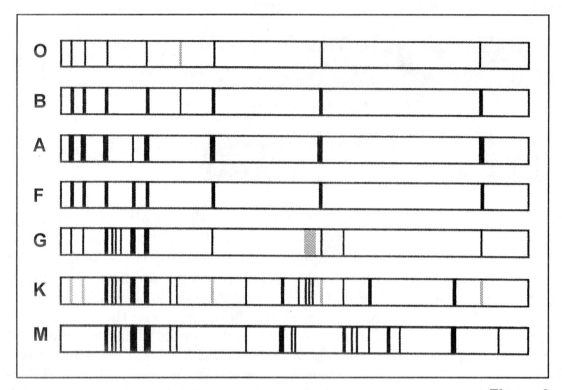

Figure 2

Spectral typing of stars can now be so accurately done that the original types are divided into ten sub-types, ranging from 0 (hottest end) to 9 (coolest end). For example, the range from Type A through Type G would include:

A0 A1 A2 A3 A4 A5 A6 A7 A8 A9 F0 F1 F2 F3 F4 F5 F6 F7 F8 F9 G0 G1 G2 G3 G4 G5 G6 G7 G8 G9

An A0 star is a little hotter than an A3 star. An A9 star is a little hotter than an F2 star. The Sun's spectral type is G2, which means that it is a moderately hot star in the Type G (yellow star) category.

In 1905, the Danish physicist Ejnar Hertzsprung first noticed that there was a relationship between the surface temperatures and the luminosities of stars. His work went largely unknown until 1913, when Henry Norris Russell (mentioned above and a much better known figure) independently re-discovered and popularized the relationship. Now, any diagram which plots luminosity against surface temperature for stars (using whatever units are appropriate, such as absolute magnitude for luminosity, and spectral type or color index for temperature) is called a *Hertzsprung-Russell diagram* or an *H-R Diagram*.

Exercise 24: The Hertzsprung-Russell Diagram

Procedure:

Examine the H-R Diagram which follows. This is not the usual way in which such a diagram is displayed (this one is "busier" looking), but every astronomer knows how to derive the information displayed here from the more simple form, which just graphs luminosity against surface temperature.

The upper and lower axes display the two different ways of estimating surface temperature: color index and spectral type. The surface temperature, in Kelvin degrees, is also given.

The left and right-hand vertical axes display two ways of displaying stars' power levels: as luminosity (in units of the Sun's luminosity) and as absolute magnitude. Notice that, while the absolute magnitude scale is **linear** (equal, <u>additive</u> increments), the luminosity scale is **logarithmic** (equal <u>multiples</u> as increments). Working with a logarithmic scale is easier than you may think; see below.

The solid curve running from upper left toward lower right is where most stars are found on any H-R Diagram: *the main sequence*. This was the correlation of stellar temperature and luminosity first identified by Hertzsprung. Our Sun is a main sequence star. Main sequence stars have wide ranges of luminosity and surface temperature, but they all have one thing in common: **they all produce energy by the fusion of hydrogen to helium in their centers**.

Stars NOT on the main sequence are "evolved;" that is, they are using (or getting ready to use) the fusion of other elements to supply their energy. Those above and to the right of the main sequence have become LARGER stars as well, and we can estimate their sizes by realizing that a star's surface temperature determines how <u>bright</u> its surface is, as well as the surface's overall color. Any G2 star will have a surface as bright as the Sun's, so the only way a G2 star can be more LUMINOUS (total power) than the Sun is if it is LARGER than the Sun. <u>Stars along any one of the diagonal dashed lines will be of the same size</u>. Those roughly 10 times the Sun's size are called giants, while those 100 times the Sun's size are supergiants and those approaching 1000 times the Sun's size are sometimes called hypergiants. At the other extreme (lower part of the diagram) are objects only a few hundredths of the Sun's size; these are white dwarfs, no longer "official" stars with internal fusion but instead the dense, collapsed remnants of stars.

The dot-dash curves, and the numerical values near the main sequence, were obtained from studies of stars in star clusters of different ages. Only the youngest clusters have powerful, blue (hot) main sequence stars in them, so we know that blue main sequence stars age much more rapidly than yellow or red ones. We also know that the hotter main sequence stars are the more massive ones. By studying hundreds of star clusters, we have been able to piece out how the physical properties (like luminosity, surface temperature and size) of a star of known mass will

Exercise 24: The Hertzsprung-Russell Diagram

CHANGE as it evolves. The dot-dash curves are known as **evolutionary tracks**. The number next to each track (such as 3M$\odot$) is the star's mass in solar masses, and the number below each is the *main sequence lifetime* in years for a star of that mass. The SUN has a mass of 1 M$\odot$, a radius of 1 R$\odot$, a luminosity of 1 L$\odot$, an absolute magnitude of +5, a color index of +0.73, and a main sequence lifetime of 10^{10} years.

You will be plotting locations of stars on this H-R Diagram and deriving some of their other physical properties. As you plot, look for <u>evolutionary connections</u> between different stars. For example, if two stars lie along the same evolutionary track but one is on the main sequence and one is a giant, *then the giant is what the main sequence star will look like LATER in its existence* (or, alternatively, the main sequence star represents what the giant looked like BEFORE it evolved).

The luminosity scale is logarithmic with a major interval of 100. It is divided into five smaller steps, each of which represents a multiple of about 2.5 . The radius and main sequence lifetime scales are logarithmic with a major interval of 10. They can be divided into five smaller steps, each of which represents a multiple of about 1.6.

A logarithmic scale with a range of 1 to 10 and five subdivisions would have values at each subdivision of approximately:

$$1 — 1.6 — 2.5 — 4 — 6 — 10$$

A logarithmic scale with a range of 1 to 100 and five subdivisions would have values at each subdivision of approximately:

$$1 — 2.5 — 6 — 16 — 40 — 100$$

If you are estimating between values in a 1 to 10 range, or in a 1 to 100 range, such as between 10^8 and 10^9 years main sequence lifetime, imagine the range as divided into five equal parts, then use the subdivision values above to find the number which multiplies the 10^8 year starting value; that is, the main sequence scale from 10^8 to 10^9 years would be subdivided as:

$$10^8 \text{ yrs} — 1.6 \times 10^8 \text{ yrs} — 2.5 \times 10^8 \text{ yrs} — 4 \times 10^8 \text{ yrs} — 6 \times 10^8 \text{ yrs} — 10^9 \text{ yrs}$$

Graph the data points given in the table that follows, and use their position on the H-R Diagram to estimate the values in regions of the table left blank. Then answer the questions below.

Questions:

Of all the stars you plotted, which one best represents what our Sun may look like in the future? Explain how you chose this star, then describe its physical properties.

Exercise 24: The Hertzsprung-Russell Diagram

The white main sequence star *Vega* is one of the brightest stars in the night sky, but being more massive than the Sun its lifetime will be considerably shorter. Of the stars you plotted, which one(s) best represent what *Vega* will look like later in its existence? Describe what it will look like!

Kruger 60B and *Betelgeuse* are both red stars, but they are located in very different regions of the H-R Diagram. Why are the two stars (so similar in surface temperature) so different in their other physical properties?

 See the LAB REPORT CHECKLIST for guidelines in preparing your report.

The Hertzsprung-Russell Diagram

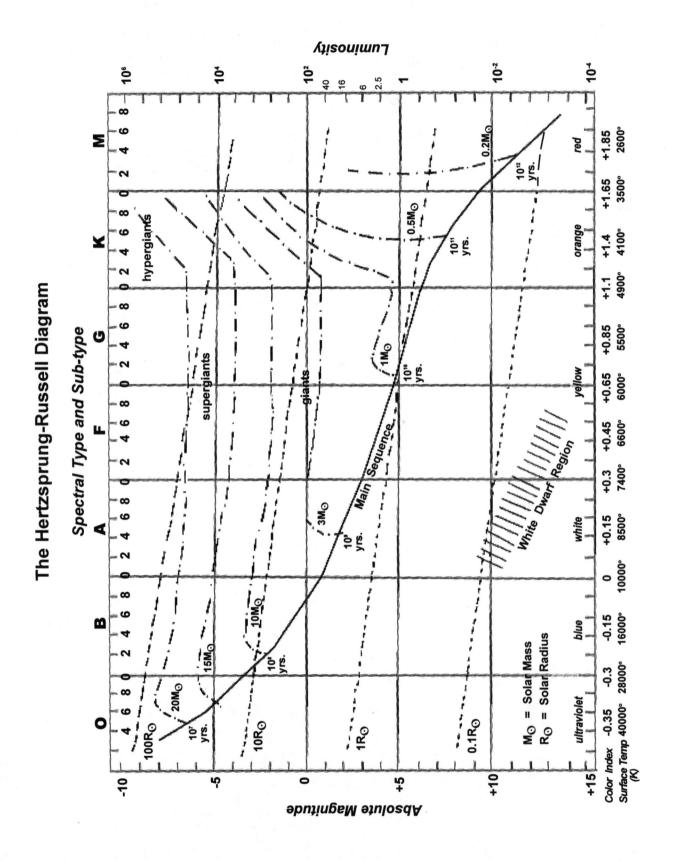

Exercise 24: The Hertzsprung-Russell Diagram

Physical Properties of Stars

Star	Absolute Magnitude	Luminosity (L⊙)	Spectral Type	Luminosity Class	Surface Temp (K)	Color Index	Mass (M⊙)	Radius (R⊙)	Time on Main Seq. (yrs)
Sun	+4.8	1	G2	main seq.	5800	+0.73	1.0	1.0	10^{10}
Rigel	+0.6	60000		supergiant		0.00			10^{7}
Vega	+0.6				10000				4×10^{8}
RR Lyrae	+0.6		F2					6	10^{9}
Procyon	+3			main seq.					
Procyon B	+13			white dwarf		+0.40	*		*
Capella		100	G6						
Pollux	+1			giant		+1.20			
Aldebaran	-1						1.0		
Betelgeuse	-6.5		M2						
Kruger 60B	+11		M4		2800				

* Value cannot be determined because of theoretical uncertainties in how the white dwarf formed

Exercise 24: The Hertzsprung-Russell Diagram

Cepheid Variable Stars I: Creating Light Curves

You Will Need: • Ruler
                      • Calculator
                      • Graph paper

Purpose: To understand how observations can be phased to yield a better representation of a variable star's brightness variations

Background:

Stars are balls of hot gas (mostly the elements hydrogen and helium), held together by their own (self) gravitation but also producing prodigious amounts of energy through nuclear reactions. Heated gases tend to expand, and that tendency to expand, called pressure, is greatest deep inside the star. Pressure drops as one goes nearer to the star's surface, and that pressure change produces LIFT for the gases, as it does for an airplane wing (see *Figure 1*).

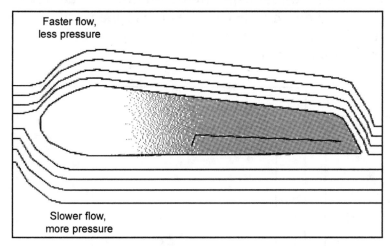

Faster flow, less pressure

Slower flow, more pressure

Figure 1

For most of its existence, any star is in a condition where, at each point inside it, the <u>weight</u> of the gases above that point is balanced against the <u>change in pressure</u> there. The balance is called *hydrostatic equilibrium* ("HSE" for short). and is an important principle that astronomers use to analyze the interiors of stars and how they evolve. The evolutionary stage of a star (such as main sequence, giant, or supergiant) sets its overall size and temperature, but HSE ensures that it will STAY at that size and temperature during that stage of existence.

However, for some stars, there comes a time when their outer layers are no longer in strict hydrostatic equilibrium; they alternately swell and shrink, sometimes in a repeatable fashion. Such stars are known as *pulsating variables*. Among the best known of the many varieties of stars which pulsate are those called *Cepheid* (SEE-fee-id) *variables*.

Cepheid variables are named after one of the nearest and best-studied examples: the star *Delta Cephei*, usually the fourth brightest star in the northern constellation CEPHEUS. *Delta Cephei* varies between apparent visual magnitudes 3.52 and 4.36 (a factor of 2.2 in brightness) over a time span of 5.3663 days.

Exercise 25: Cepheid Variable Stars I: Creating Light Curves

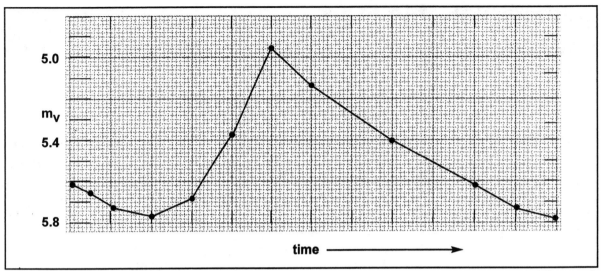

Figure 2

By comparing the magnitude of *Delta* to those of nearby, non-varying stars on any night and doing this over many nights, you would begin to understand the details of the star's light changes. A graph of visual magnitude ("m_V") versus time is called the LIGHT CURVE of a variable star. *See Figure 2.* It represents a typical light curve for a star with this approximate pulsation period.

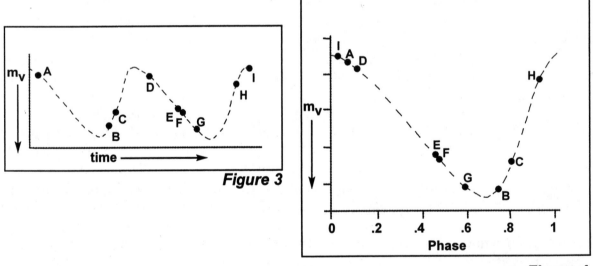

Figure 3

Figure 4

Because ground-based observatories are often bothered by bad weather, it is often difficult to observe a Cepheid's brightness changes over one single pulsation cycle. Instead, observations are usually spread over many cycles (see *Figure 3*). Fortunately for the people who study them, the pulsation of Cepheids is very regular and repeatable, to better than a few parts per million! Since each new cycle is virtually identical to the previous ones, it is possible to piece together a

Exercise 25: Cepheid Variable Stars I: Creating Light Curves

representative light curve (see *Figure 4*) from the different cycles, by a technique called *phasing*. Even for stars where the pulsation is not as repeatable, this technique allows astronomers to estimate a representative AVERAGE length of the pulsation over many cycles.

Procedure:

To phase observations into a light curve, we must know the amount of time needed for one pulsation cycle (the *period*, P), and we must have observed (or estimated) a particular time (tMax) when the star was at MAXIMUM brightness. If we have these, and the times (t) of all other observations, then the phase ϕ of a particular observation will be:

$$\text{(If } t \geq tMax) \qquad \phi = \text{the decimal part of } \frac{(t - tMax)}{P}$$

$$\text{(If } t < tMax) \qquad \phi = 1 \text{ } plus \text{ the decimal part of } \frac{(t - tMax)}{P} \text{ }*$$

* Only true if t< tMax (the decimal part will have a negative value)

The table on the following page lists visual apparent magnitudes (obtained with a photometer) and times of observation (in days elapsed since the first observation) for a Cepheid variable called *VV Cassiopeiae*, which has a pulsation period of 6.21 days. Since the observations span many more than 6.21 days, we know that they span several pulsation cycles of VV Cas.

Complete the table, using the phasing technique and drawing upon some of the table entries to check your calculations.

On graph paper with 10 squares to the centimeter or 20 squares to the inch, lay out a vertical axis (ordinate) which runs from 10.1 at the <u>top</u> to 11.3 at the <u>bottom</u>**, making a tick mark for every 0.1 or 0.2 change in value. Label this axis "m_v" (apparent visual magnitude). Label the horizontal axis (the abscissa) "phase"; this should run from 0.0 (on the left) to 1.0. Plot the values of m_v and phase for VV Cas on the graph. When you have plotted all the points, connect them with a curve that follows their location <u>closely</u>. These points are very accurate, at least if you did the phases right!

**   This is backwards from the usual graphing technique, but <u>lower</u> magnitudes represent <u>greater</u> brightness, which we want at the top.

Exercise 25: Cepheid Variable Stars I: Creating Light Curves

Observations of VV Cas (P = 6.21 days)

m_V	t	t – tMax	(t-tMax) / P	Phase, φ
10.63	0.00	-13.92	-2.08	0.92
10.40	1.05	-11.87	-1.91	0.09
11.01	10.05	-2.87		
10.32	12.92	0.00	0.00	0.00
10.59	13.98	1.06	0.17	0.17
10.76	14.99	2.07	0.33	0.33
10.89	15.93	3.01		
11.03	16.85			
11.11	17.93	5.01	0.81	0.81
10.35	18.93	6.01	0.97	0.97
10.49	19.93	7.01	1.13	0.13
10.70	20.92			
10.84	21.91	8.99		
11.18	23.89			
10.89	36.95	24.03		
11.08	48.98	36.06	5.81	
10.42	56.88	43.96	7.08	0.08
11.00	59.81	46.89		
10.77	61.89	48.97	7.89	0.89

Questions:

What percentage of its pulsation cycle does VV Cas spend in brightening (going from large m_V to small m_V)?

You derived the average light curve for VV Cas by *assuming* that its pulsation period was 6.21 days. What would be the effect on the light curve of assuming a different period, say 6.00 days?

 See the LAB REPORT CHECKLIST for guidelines in preparing your report.

Exercise 25: Cepheid Variable Stars I: Creating Light Curves

Cepheid Variable Stars II: Investigating Pulsation

You Will Need: • Ruler
  • Calculator
  • Graph paper

Purpose: To examine the internal structure of variable stars through observations of their pulsation

Background:

The regular pulsation that Cepheids display does not occur in stars of every surface temperature or power level (luminosity). Any plot of surface temperature versus luminosity for stars is called a Hertzsprung-Russell or H-R Diagram* after the two men who developed it. Cepheid variables are only found in a narrow region on the H-R diagram, called the "instability strip."

A. The Instability Strip

The table below lists the properties of some typical Cepheids: their pulsation periods in days, their average spectral types, and their *average absolute visual magnitudes*. Unlike apparent magnitude, which ranks the energy received from a star, absolute magnitude ranks the luminosity of a star. Spectral type and luminosity both change during pulsation, so averages must be used.

Properties of Typical Cepheids

Name **	P (days)	Avg. Spect. Type	Avg. M_v
TT Aquilae	13.755	G0	-4.5
RT Aurigae	3.728	F7	-3.1
RX Aurigae	11.624	F9	-4.5
SU Cassiopeiae	1.949	F6	-2.4
Delta Cephei	5.366	F8	-3.5
X Cygni	16.387	G2	-4.5
SU Cygni	3.846	F6	-3.2
Zeta Geminorum	10.151	G0	-4.2
T Momocerotis	27.025	G4	-5.4
Y Ophiuchi	17.124	G1	-4.8
RS Puppis	41.388	G3	-6.1
U Sagittarii	6.745	F8	-3.9
WZ Sagittarii	21.850	G5	-5.1
SV Vulpeculae	45.012	G4	-6.1
S Vulpeculae	68.46	G6	-6.7

* See the exercise called "The Hertzsprung-Russell Diagram for further details."

Exercise 26: Cepheid Variable Stars II: Investigating Pulsation

Procedure:

Figure 1 on the following page shows a portion of the H-R diagram. The main sequence, and regions where giant stars, supergiant stars, and luminous supergiant stars are found, are indicated. Photocopy *Figure 1*, then plot on it the values of average M_v, and average spectral type for the stars in the table.

Use different plot symbols for stars on different ranges of period values:

> For P < 5 days, use open circles (O)
> For P = 5 days up to P=10.9 days, use filled circles (●)
> For P = 11 days up to P=20 days, use crosses (+)
> For P > 20 days, use diagonal crosses (X)

** Variable star names contain the constellation name in the Latin genitive (possessive) form (e.g., WZ Sagittarii means, "variable star WZ of *Sagittarius*) and a letter or letter-number designation, which generally follows the date of discovery as a variable star. The notation is a little strange, starting at R to Z, then RR to RZ, SS to SZ...to ZZ, then AA to ZZ, and so on up to QQ to QZ, except that no J's are used (This is because J could be confused with I). This names the first 334 variables in a constellation. If more than 334 variables are found, the rest are called V335, V36...Some constellations that lie along the Milky Way (such as *Sagittarius*) can contain thousands of variable stars!

Exercise 26: Cepheid Variable Stars II: Investigating Pulsation

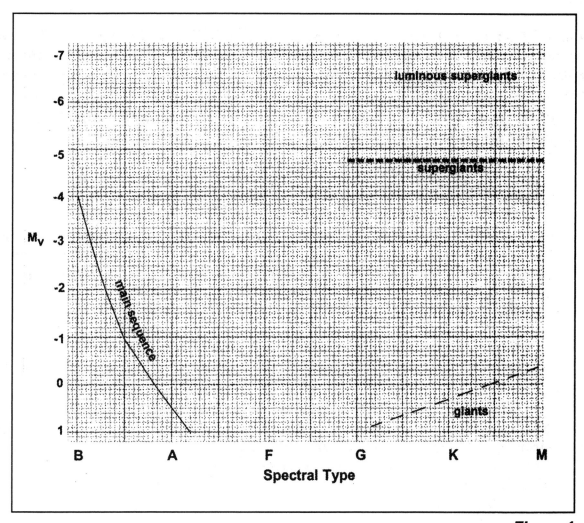

Figure 1

Questions:

What kinds of stars are Cepheid variables? What does this tell you about their evolutionary state?

The Cepheid variable X Cygni has an average spectral type (G2) like the Sun, but while the Sun's absolute magnitude is about +4.8, the absolute magnitude of X Cygni is -4.5!

a) How many magnitudes more luminous than the Sun is X Cygni? Express the result as a positive number.

b) If that number is N, then the luminosities of the two stars compare in the proportion:

Exercise 26: Cepheid Variable Stars II: Investigating Pulsation

$$\frac{L_{X\ Cyg}}{L_{Sun}} = (2.512)^N \qquad\qquad \text{Calculate the amount!}$$

c) If the two stars have the same spectral type, they have the same surface temperature. Therefore, the only reason they would differ in luminosity is that they differ in SIZE. X Cygni has more surface area than the Sun. Surface area goes in proportion to the star's radius squared, so:

$$\frac{L_{X\ Cyg}}{L_{Sun}} = \frac{R_{X\ Cyg}^2}{R_{Sun}^2} \qquad\qquad$$ On average, <u>how many times</u> larger in RADIUS than the Sun is X Cygni?

You plotted the different period ranges of Cepheids with different symbols so that you could look for correlations between the star's different properties. What sorts of correlations do you find among the different quantities displayed on the graph?

B. Cepheid Pulsation

Figure 2 combines three different graphs for the star *Delta Cephei*. Its apparent visual magnitude, surface temperature, and radial velocity are all shown varying with time. Because the graphs are all plotted with the same timescale, you can directly compare the star's brightness level, temperature, and atmospheric motion at any given time. Radial velocities - motions of the star's atmosphere toward or away from the observer - are gotten from the tiny position shifts of dark lines in the star's spectra. Radial velocities are negative when the atmosphere is expanding outward, positive when it is contracting, and zero when it has momentarily halted, about to change direction.

Examine *Figure 2* and answer the following questions:

What is the apparent visual magnitude of *Delta Cep* (e.g., hottest, coolest, average) when it is brightest? When it is faintest?

What can you say about the surface temperature of *Delta Cep* when it is brightest? When it is faintest?

From its radial velocity curve, you can determine when *Delta Cep* will be largest or smallest in size. Cepheid pulsation is alternate swelling and shrinking of a star's outer layers. Imagine the star as swelling. Its largest size will NOT be reached when it is expanding <u>fastest</u>, but rather when it has CEASED swelling and is about to collapse again. That is, Delta Cep is LARGEST when its radial velocity becomes ZERO after having had negative values. Locate this point on *Figure 2*.

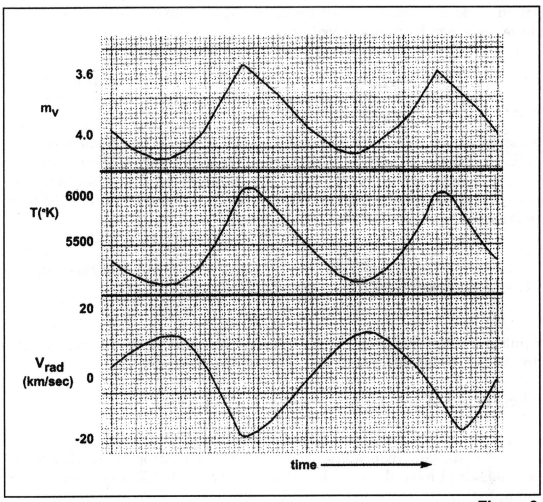

Figure 2

Where on the radial velocity curve is *Delta Cep* smallest? Locate this point on *Figure 2*.

What is its atmosphere doing (what sort of motion) when *Delta Cep* is at its brightest? Is *Delta Cep* brightest when it is largest?

How bright is *Delta Cep* when its atmosphere is falling inward (contracting) most rapidly?

Compare surface temperatures of *Delta Cep* when it is largest and when it is smallest.

A Cepheid changes both in size and temperature as it pulsates, and both of these will affect its brightness changes. Hotter stars have brighter surfaces, while larger stars have more surface area from which to radiate light.

The surface brightness of a star is proportional to the fourth power of its surface temperature (Stefan's law). *Delta Cep* ranges in surface temperature over its

Exercise 26: Cepheid Variable Stars II: Investigating Pulsation

pulsation cycle from about 5100° Kelvin to 6500° Kelvin, so the proportion of its brightnesses when hottest and coolest is:

$$\frac{B_{6500K}}{B_{5100K}} = \frac{(6500)^4}{(5100)^4}$$ Calculate the value of this proportion.

By a careful study of *Delta Cep's* radial velocity changes, astronomers can say that its radius varies from about 42 to 52 times the Sun's. If only SIZE were changed, the star's brightness would vary with the changes in its total surface area, proportional to the radius squared. Therefore, the proportion of its total brightness when largest versus smallest will be:

$$\frac{B_{largest}}{B_{smallest}} = \frac{(52)^2}{(42)^2}$$ Calculate the value of this proportion.

If a pulsating variable star can increase in brightness by getting larger, or by getting hotter, which effect is more important for *Delta Cephei*?

C. The Theory of Pulsation

We have seen that Cepheids are bright and hot as they expand, and that they take less time to increase in brightness than they do to drop in brightness. These observations lead us to the idea that Cepheids are FORCED to expand under great pressure. This pressure subsequently leaks away, and the star's atmosphere (somewhat more slowly) contracts. The pressure waves that push a Cepheid's atmosphere outward could arise from convection in the atmosphere, or from the small, nearly random motions of gas always present in a star's outer layers.

Getting the atmosphere to become unstable is another matter. When a normal gas is compressed, its atoms fly around faster and faster and its ability to absorb energy is hindered: the gas becomes more TRANSPARENT. If the gas in a star's atmosphere readily leaks away energy, it will be difficult to store up enough in the atmosphere to push the atmosphere outward. Most stars are stable (do not pulsate) because of this behavior of the gas.

The solution - how to get a gas to *store* energy when compressed rather than *leak* it - is to partially IONIZE the gas; that is, heat it until its atoms start to lose some electrons. When energy is poured into a gas that is partially ionized, it is used up in ionizing the remaining atoms of the gas; the energy is stored in, rather than leaked out of, this region of the star. When the pulsational disturbance has passed, the freed electrons recombine with their ions and release the energy, allowing the star to shrink again. Regions where hydrogen (H) and helium (He) are partially ionized

Exercise 26: Cepheid Variable Stars II: Investigating Pulsation

seem to be the best candidates for helping maintain pulsation in Cepheids. These two zones are thought to occur at depths in stars where the temperature has reached about 10,000°K and 40,000°K, respectively.

Questions:

How deep (in general terms) would you have to go to find these regions in an A-type star? In a M-type star?

Can you explain why these regions will NOT be very effective at pushing for a very hot or very cool star?

Now we see, in principle, why an instability strip exists. Other factors, like the presence of a very deep, convective atmosphere or the star's loss of matter through a stellar wind, can further affect where in the H-R diagram pulsation occurs.

 See the LAB REPORT CHECKLIST for guidelines in preparing your report.

Spiral Structure In The Milky Way

You Will Need: · A straight edge ruler

Purposes: To determine what kinds of objects are most suitable for mapping our galaxy's spiral structure, and to examine the structure of spiral arms near the Sun

Background:

You can view our galaxy, called the Milky Way, on any clear, moonless night in locations well away from city lights. All the individual stars you see in the night sky belong to our galaxy; but, stars which are so distant that they cannot be seen as individual points still reveal their presence as a hazy band of light extending across the sky.

The "band"- like appearance of the Milky Way means that the stars in our galaxy are arranged in a disk shape. The Sun is located somewhere inside the disk but not near an edge, because the band stretches completely around the sky. Surrounding our galaxy's disk is a more spherical region of space called the halo, which contains mostly older stars.

You can locate stars and other objects in the Milky Way galaxy (disk or halo regions) by using a coordinate system similar to the latitude and longitude system used on Earth. Imagine a circle running along the center of the Milky Way band; this is called the galactic equator, or sometimes the "galactic plane." You measure an angle called galactic latitude (b) in degrees above or below the galactic equator. A star on the galactic equator has b = 0°, while one below it has a negative value of b and one above it has a positive value. Galactic latitudes range from +90° to -90°.

The Milky Way is strongest and widest low in the southern sky on late summer evenings, most especially around the constellations Scorpius and Sagittarius, and about 8,500 parsecs* away from us in that direction is the center of our galaxy. You measure an angle called galactic longitude (ℓ) counterclockwise along the galactic equator starting from the direction to the galactic center. Galactic longitudes range from 0° to 360°. See *Figure 1*.

---

*   The *parsec* is a useful unit of distance when studying stars and galaxies. One parsec equals about 19 million million miles or 31 million million kilometers. Our nearest neighboring star system (Alpha Centauri) is about 1.3 parsecs or 25 million million miles or 40 million million kilometers from us. In galactic structure work, the *kiloparsec* is also used; one kiloparsec (kpc) equals 1000 parsecs.

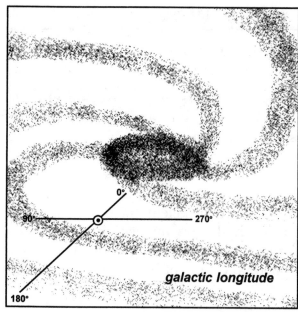

galactic longitude

Figure 1

For example, a star having $\ell = 0°$ and $b = 0°$ would be seen along the galactic equator in the direction toward the galactic center. A star with $\ell = 180°$ and $b = +30°$ will be seen above the galactic plane in the direction facing away from the galactic center.

Astronomers have discovered that stars at different distances from the galactic center move around it at different speeds; this *differential rotation* of the galaxy's disk means that the Milky Way's beautiful spiral arms cannot be permanent, solid structures. They are more like moving "traffic jams" where stars and clouds of gas and dust are temporarily piling up. What causes them to pile up in certain areas? Ripples in the overall gravitational pull of the disk, which together are called the **spiral pattern**, may have been created by past close approaches of other galaxies, and it is there that such pileups or **density waves** of material will occur and outline what we call the spiral arms.

Star formation also occurs in the spiral arms. See *Figure 2*. As clouds of gas and dust orbit around the galactic center, they run into the spiral pattern and are slowed and compressed. This starts star formation! The spiral arms are home to star clusters and longer, more straggly chains of young stars called associations. The most obvious young objects are the powerful, blue stars. But, in principle, any object known to be young, or known to have a short lifetime, can be used to map out where the spiral arms in a galaxy are. Such objects are called **spiral tracers**.

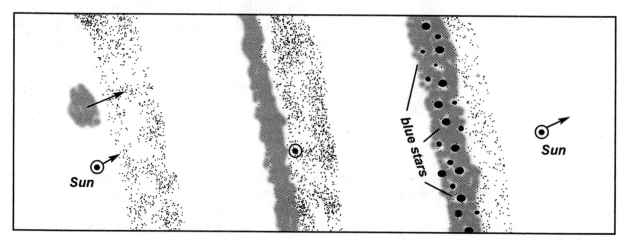

Sun

blue stars

Sun

Figure 2

Exercise 27: Spiral Structure In The Milky Way

Old or long-lived objects, like our Sun, do not make very good spiral tracers. Although they may also get caught for awhile in the "traffic jam," they exist long enough to emerge from the other side of the density wave region and move away (as shown in Figure 2). Also, the older a star is, the more time it has to move up or down out of the galactic plane, spreading slowly into the galaxy's outer region or halo, Older stars are less connected to the spiral pattern because it is mostly developed in the disk.

Procedure:

The first map is an equal-area projection of the entire sky, using galactic coordinates (b, ℓ). The horizontal line running across the middle of the map is the galactic equator, with the $\ell = 0°$ (Sagittarius) direction at the center.

Table 1 contains galactic latitude and longitude information, and distances in kiloparsecs, for some groups of hot, young stars called **OB associations**. OB associations contain hundreds of stars. They form whenever clouds of dust and gas enter part of the spiral pattern. We know these associations are young because the stars in them (blue stars of spectral types O and B) have very short lives (millions of years or less).

Plot the positions of these groups on the first map.

Now examine *Table 2*. This lists galactic coordinates for some globular star clusters, large groupings of tens to hundreds of thousands of yellow and red stars. **Using a different plot symbol or a different symbol color, plot the positions of these stars on the same map.**

{If you have access to planetarium software such as Voyager II or III which allows you to select stars by spectral type, choose only stars of Type O (all luminosity classes) to plot. In Voyager, this is found as the Spectral Types choice under Star Selection in the pull-down Chart menu. Turn on the Milky Way outline from the pull-down Display menu, and turn on the galactic equator from the Galactic Equator choice in that same menu. Select a deep (large number) magnitude limit, also in the Chart menu. Choose a wide (120° - 180°) field of view (zoom in or out, then look at Chart Panel in the Format window), and display the data. Notice the locations of the O stars relative to the galactic plane. Make a second display, but this time choose stars of spectral type M and of luminosity class III (giants). If possible, print out both displays.}

Exercise 27: Spiral Structure In The Milky Way

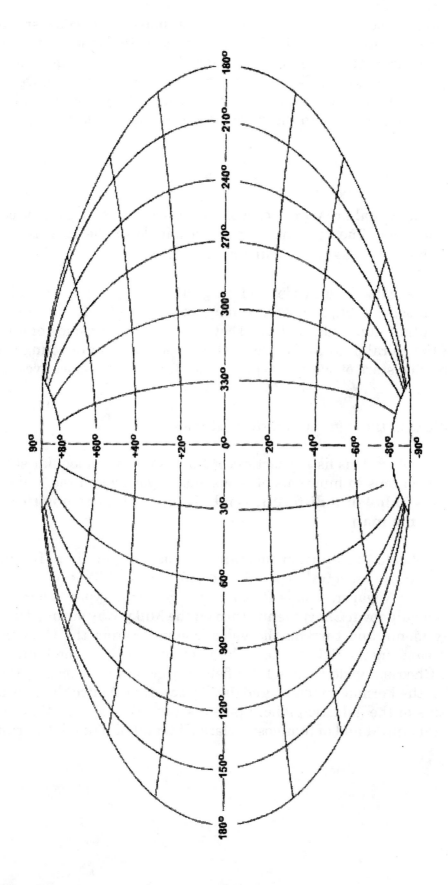

Map 1

Exercise 27: Spiral Structure In The Milky Way

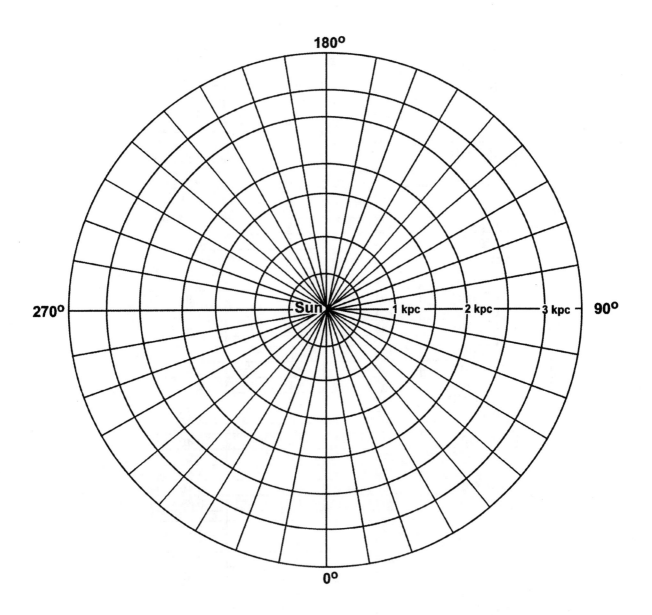

Map 2

Exercise 27: Spiral Structure In The Milky Way

Table 1
OB Associations

Group Name **	Longitude (ℓ), degrees	Latitude (b), degrees	Distance (kpc)
Sgr OB1	8	-1	1.6
Sgr OB4	12	-1	2.4
Ser OB1	17	0	2.2
Ser OB2	19	+1	2.0
Vul OB1	60	+1	2.0
Cyg OB3	72	+2	2.3
Cyg OB1	76	+1	1.8
Cyg OB9	79	+2	1.2
Cyg OB2	81	+1	1.8
Cyg OB4	84	-8	1.0
Cyg OB7	90	+3	0.83
Lac OB1	98	-17	0.60
Cep OB2	103	+5	0.83
Cep OB1	104	-2	3.5
Cep OB3	112	+5	0.87
Cas OB5	117	-2	2.5
Cep OB4	119	+6	0.84
Cas OB4	121	+1	2.9
Cas OB14	121	+2	1.1
Cas OB1	125	-1	2.5
Cas OB8	130	-1	2.9
Per OB1	135	-4	2.3
Cas OB6	137	+2	2.2
Cam OB1	143	+2	1.0
Per OB3	148	-5	0.17
Per OB2	159	-17	0.40
Aur OB2	172	+1	3.2
Gem OB1	190	+2	1.5
Ori OB1	206	-18	0.46
Mon OB1	203	+1	0.55
Mon OB2	208	-1	1.5
CMa OB1	226	-1	1.3
Pup OB1	244	+1	2.5
Vel OB2	263	-7	0.46
Vel OB1	267	-1	1.4
Car OB1	288	0	2.5
Car OB2	290	0	2.0
Cen OB2	295	-1	2.5
Cen OB1	305	0	2.5
Nor OB1	328	-2	2.5

** The name of the association contains the constellation in which it is located, "OB" for OB association, and a number indicating order of discovery. For example, "Ser OB2" is the second OB association discovered in Serpens.

Exercise 27: Spiral Structure In The Milky Way

Table 2
Globular Star Clusters

Cluster Name ***	Longitude (ℓ), degrees	Latitude (b), degrees	Distance (kpc)
47 Tuc	306	-45	4.6
NGC 1851	244	-35	10.8
NGC 3808	283	-11	9.2
ω Cen	309	+15	5.2
Messier 3	42	+78	10.0
Messier 5	4	+47	7.2
Messier 4	351	+16	2.4
Messier 13	59	+41	7.2
Messier 12	15	+26	5.4
Messier 10	15	+23	4.4
Messier 92	68	+35	7.8
NGC 6397	339	-12	2.4
NGC 6541	349	-11	6.8
Messier 22	9	-8	3.0
NGC 6723	0	-18	9.0
NGC 6752	337	-26	4.3
Messier 55	9	-24	5.7
Messier 15	65	-27	10.1
Messier 2	54	-36	11.2
NGC288	147	-89	8.7
Messier 68	299	+37	9.6
Messier 53	333	+80	17.2
Messier 14	21	+14	9.9
Palomar 2	171	-9	13.6
NGC 3201	277	+9	4.4

***   Cluster names reflect their discoverers (such as Charles Messier or astronomers at Palomar Observatory), sequence numbers in catalogs (NGC = New General Catalog), or the cluster's overall glow being mistaken by the unaided eye for a single bright star (such as, 47 Tuc or ω Cen).

Exercise 27: Spiral Structure In The Milky Way

You may have noticed, in comparing *Tables 1* and *2*, that the globular star clusters listed are much more distant than the OB associations. That's because OB associations are mostly found near the galactic plane, and globular clusters are usually not. Particles of interstellar dust, which is mostly found in the disk, cause a "haze" which prevents visible light reaching us from very great distances, but it doesn't much affect our views up (or down) out of the disk where the globular star clusters are found. Robert Trumpler was one of the first people to demonstrate (around 1930) that our view along the galactic plane was dimmed by interstellar material; the more distant the object, the more dust gets in the way and the greater the dimming.

If you examine your plots of star groupings on the first map, you will notice that OB associations are spread fairly evenly along the galactic plane but globular star clusters show a preference for the center region (around $\ell = 0°$) of the map. If the Sun were at the center of our galaxy, stars, star clusters and other objects should be found in roughly equal numbers in all directions around us, so the location of globular clusters seems unusual. In 1917, astronomer Harlow Shapley determined the distances to globular star clusters and from them an estimate of the distance to the galaxy's center.

The second map shows how the regions of our galaxy near the Sun would look if viewed from above the galactic plane. The coordinates of this polar map are galactic longitude ℓ (marked around the rim of the largest circle) and distance from the Sun in kiloparsecs (marked on every second circle, outward from the center).

On the polar map, carefully plot the longitudes and distances of the OB associations from *Table 1*.

Questions:

If you used planetarium software, estimate the spread (in degrees, above and below the galactic plane) of the O stars. Compare their overall location in relation to the galactic plane with that of the M giant stars. Can you explain why there should be a difference? (Use information within this exercise and from your own knowledge of stars and how they evolve.)

On the first map, compare the locations relative to the galactic plane of OB associations and globular star clusters. Why should they be so different?

Can you see the locations of some spiral arms from the positions of OB associations that you plotted on the second map? Roughly sketch in "arm outlines" along groups of points. (Hint: plotting more associations shows that the arms are tilted down a bit toward the right in the graph. See *Figure 3*.)

Exercise 27: Spiral Structure In The Milky Way

How much uncertainty is there in where the arms are located? Is our Sun in or near a spiral arm?

The Milky Way is most noticeable (brightest) in directions where you are looking ALONG a spiral arm, as shown in *Figure 3*. Directions A, B, C and D are all in the galactic plane. Since very few stars are found between the arms, an observer looking in direction A will see virtually no Milky Way glow. Direction B will look brighter, since it looks across an arm and intersects the locations of some stars, and direction D will be brighter still because it crosses the stars in TWO arms (one of them nearby). But direction C will show the brightest Milky Way because the observer looks along the direction of most stars.

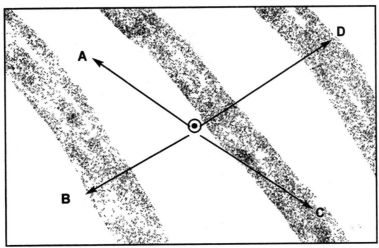

Figure 3

The following list gives the approximate galactic longitudes of some constellations which lie along the galactic plane:

Sagittarius -- 0°	Scutum -- 30°	Cygnus -- 70°
Cassiopeia -- 120°	Perseus -- 150°	Auriga -- 170°
Gemini -- 190°	Canis Major -- 230°	Puppis -- 250°
Crux -- 300°	Norma -- 330°	Scorpius -- 350°

In, or between, which of the listed constellations will the arms you traced on the second map be most noticeable to observers? Justify your answers! (Remember, too: the more distant the star, the more its light is dimmed by interstellar dust.)

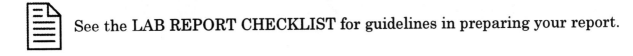 See the LAB REPORT CHECKLIST for guidelines in preparing your report.

Exercise 27: Spiral Structure In The Milky Way

Astronomy on the Internet

You Will Need: • A PC or Macintosh which has access to the internet
                       • A Web browser, such as *Netscape* or *Internet Explorer*

Purposes: To learn internet basics, to view some popular and useful astronomy sites on the World Wide Web and follow links to other sites, and to use search engines to find current information about astronomy topics

Background:

Networking is using one computer to link to others, so that information can be transferred among them.

The term "internet" refers to a collection of programs on a variety of computers at host institutions around the world. These programs allow you access to the computers and, through them, access to still others. As some of you may already know, the internet had its beginnings in the early 1970s as an application called ARPAnet, developed by the U.S. Department of Defense as a means of maintaining a flow of information even if one or more computers in the network were disabled (such as by a nuclear strike). In the late 1980s, the National Science Foundation networked five supercomputer centers, forming NSFnet. Gradually, the business potentialities of networking were realized and developed; and, with the growth of the personal computer industry and the production of ever-faster microprocessor chips, the internet has become available to schools, businesses and households around the world.

The first internet programs were primarily text only, used for searching libraries, catalogs and databases, and many of them still exist. But with the development of **hypertext**, a command language which allowed users to create their own *multimedia* sites, the internet finally came of age. Much of what the general public knows about the internet is through its multimedia, hypertext sites, which collectively are called the "World Wide Web."

Every location on the Web is called a **Web site** and has its own designation, called its **URL** ("you-are-ell," not "earl;" uniform resource locator) or simply "Web address." I'm sure you're familiar with some of these, such as

www.almanac.com   (the *Old Farmer's Almanac*)
www.nasa.gov   (National Aeronautics and Space Administration)
www.cnn.com   (Cable News Network)

The different parts of the address are separated by periods, and the last portion of

Exercise 28: Astronomy on the Internet

the address refers to the type of institution where the computer that stores the information is found. Some addresses are even more specific, with other words following the main address and separated from it by slashes (/).

When you want to go to a particular Web site, you have to specify its URL and also (preceding it) type the command "http://" (http = 'hypertext transfer protocol'), so that the access program you are using on your computer knows it is a search request.

If you don't work at a place where a means of 'surfing the Net' is already on your computer, you can purchase access through one of a number of **Internet access providers** or **on-line services**, such as America Online. These allow you access to World Wide Web addresses and also provide other services, such as "hot pick" Web site choices and the sending and displaying of e-mail messages. The providers and services also allow you access to programs called **Web browsers**, such as *Netscape* and *Internet Explorer*, **catalog sites** such as *Yahoo!* and *Northern Light*, and **search engines**, such as *Alta Vista*, *Lycos* or *Webcrawler*, which have more complete address listings of topics than some of the catalog sites.

If you are using the computers at your school rather than your home computer, you will probably be using either *Netscape* or *Internet Explorer* (whichever your computer system has loaded) to explore sites on the World Wide Web. If you are working from home and using (for example) AOL as an on-line service, you can still go to one of those web browsers by opening their home pages, such as **netscape.com**.

The first display you see when bringing up these programs (or any others on the Web) is called the **home page**. It has a *menu bar* at the top, with **pull-down menus** that give you options such as downloading and printing files, conducting topic searches, and making bookmarks (lists of URLs you may want to examine again). Below the pull-down menus are **icon buttons** that allow you to go backward or forward among files you've downloaded, stop the loading of a new file (such as one which might take hours to transmit!), and other useful things. Below the menu bar is an **address** or **location field** (a long, narrow window) where you can type in the URL of a site you want to explore. Hitting your computer's ENTER or RETURN key, or clicking on a GO button if there is one, will take you to that site. See *Figure 1*.

Used with permission of YAHOO!

Figure 1

Exercise 28: Astronomy on the Internet

Within the text of most home pages (and on any other pages which you may open), are words which are underlined and shown in color (often blue). These are **links** to other Web addresses of related interest. Position your computer cursor (the arrow or pointing hand) over a link; its URL will be displayed at the bottom of the screen if you are using either Netscape or Internet Explorer. Now click on your mouse to activate the link and bring up the new page.   See *Figure 2*.

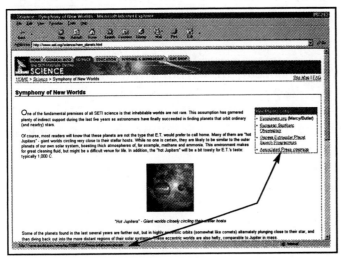

Used with permission of SETI Institute

Figure 2

Service providers and Web browsers give you access to URLs in general areas of interest, such as travel, science or sports. They also all allow you to do searches for more detailed topics by typing in key words. **But, be careful!** Unless you are somewhat restrictive in your choice of a topic, you may be informed that the browser has identified thousands (or more) of URLs that might match it! For example, if you searched for Web sites about 'War,' I can't even imagine how many matches you'd get! If you typed 'Civil War,' you would still get a huge number of matches. However, if you typed 'Civil War AND Battle of Spotsylvania,' the Web sites containing both topics areas would be a much smaller number. This is called doing a **Boolean search** (after the 19th Century mathematician George Boole).

Procedure:

Look at the following astronomical Web sites. Most sites have URLs which include "www", but some will not. <u>Write a brief description</u> of what each home page contains, and <u>try to download and print</u> (if possible) any picture or text files of interest at that site.

> www.stsci.edu
> www.skypub.com
> www.astronomy.com
> antwrp.gsfc.nasa.gov/apod/astropix.html
> www.seti.org

Pick ONE of the URLs above and (starting from the home page) follow ONE LINK on that page. When you reach the new page, follow ONE LINK on that one. <u>Keep a record of the URL</u> of each new link, and <u>jot down a brief description</u> of what each new page displays for you. At any time in your linking, you can always

Exercise 28: *Astronomy on the Internet*

return to the previous page by clicking on the Back button. **Follow at least five links to other pages.**

Now open a **search engine**, such as *Webcrawler, Lycos,* or *AltaVista.* Their URLs are just "www" plus their names "dot-com," like www.webcrawler.com. **Do a search on ONE of the following questions**, and (assuming your search doesn't fail!) write a brief description of what you found. You may also attach any text or pictures that you found and printed during your search.

When does the next *total eclipse of the Sun* occur, and from what locations can it be seen?

What bright *planets* will be visible in your evening sky tonight, this week or this month?

Are any bright *comets* currently visible?

What do we know about the object **PSR 1937+21**?

What discoveries in astronomy were announced during the past week?

What are the latest findings about Jupiter's moon **Europa**?

(Hint: for current events, you may also want to look at some of the news sites on the Web.)

 See the LAB REPORT CHECKLIST for guidelines in preparing your lab report.

Exercise 28: Astronomy on the Internet

Lab Report Checklist

✔   **Title** of the experiment, your **name**, and the **date** the report is submitted

✔   **Description of equipment and observations**
   · Brief summary of the purpose
   · Details about the equipment and how it was constructed
   · When/how the observations were taken (and of what objects)
   · Identification of <u>variables</u> and <u>controls</u>
   · What kinds of calculations (when applicable) were needed

✔   **Data and results**
   · Tables and/or graphs, sometimes diagrams
   · Specific information about location, weather, time of observation, etc.
   · Evaluating the level of UNCERTAINTY
   · Average deviation or σ when appropriate
   · <u>Where</u> the greatest difficulties in obtaining accurate information originated

✔   **Conclusions**
   · What concepts or relationships were demonstrated to you?
   · How could the exercise be made more <u>accurate</u> or more <u>general</u>?

Index

For topics with more than one page number listed, numbers in <u>underline</u> indicate primary discussion of the topic (on that page and sometimes on following pages). Those in **boldface** have an illustration/illustrations of the topic. Some topics will have <u>**both**</u>.

Index